Effective and Robust Gas Turbines

Effective and Robust Gas Turbines

Edited by **Eugene Bradley**

New York

Published by NY Research Press,
23 West, 55th Street, Suite 816,
New York, NY 10019, USA
www.nyresearchpress.com

Effective and Robust Gas Turbines
Edited by Eugene Bradley

© 2015 NY Research Press

International Standard Book Number: 978-1-63238-120-0 (Hardback)

Printed in the United States of America.

Contents

Preface   VII

Introductory Chapter   **Overview**   1
Konstantin Volkov

Chapter 1   **Ultra Micro Gas Turbines**   5
Roberto Capata

Chapter 2   **Energy, Exergy and Thermoeconomics Analysis of Water Chiller Cooler for Gas Turbines Intake Air Cooling**   51
Rahim K. Jassim, Majed M. Alhazmy and Galal M. Zaki

Chapter 3   **Energy and Exergy Analysis of Reverse Brayton Refrigerator for Gas Turbine Power Boosting**   77
Rahim K. Jassim, Majed M. Alhazmy and Galal M. Zaki

Chapter 4   **The Selection of Materials for Marine Gas Turbine Engines**   101
I. Gurrappa, I. V. S. Yashwanth and A. K. Gogia

Chapter 5   **Gas Turbines in Unconventional Applications**   121
Jarosław Milewski, Krzysztof Badyda and Andrzej Miller

Chapter 6   **Gas Turbine Diagnostics**   165
Igor Loboda

Chapter 7   **The Recovery of Exhaust Heat from Gas Turbines**   187
Roberto Carapellucci and Lorena Giordano

Chapter 8   **Models for Training on a Gas Turbine Power Plant**   213
Edgardo J. Roldán-Villasana and Yadira Mendoza-Alegría

Permissions

List of Contributors

Preface

The purpose of the book is to provide a glimpse into the dynamics and to present opinions and studies of some of the scientists engaged in the development of new ideas in the field from very different standpoints. This book will prove useful to students and researchers owing to its high content quality.

A descriptive account on the efficiency as well as robustness of gas turbines has been provided in this book. Various topics regarding the evaluation of gas turbines and their engineering operations are presented in the book. Methodical and experimental procedures are utilized to classify failures and quantify operating circumstances and efficiency of gas turbines. Some topics included in this book are gas turbine engine flaws, diagnosis and condition monitoring systems, operating conditions of open gas turbines, reduction of jet mixing noise, recovery of exhaust heat from gas turbines, apt substances and coatings, ultra micro gas turbines and operations of gas turbines. The exchange of scientific outcomes and ideas in this book will result in an advanced reliability on gas turbines.

At the end, I would like to appreciate all the efforts made by the authors in completing their chapters professionally. I express my deepest gratitude to all of them for contributing to this book by sharing their valuable works. A special thanks to my family and friends for their constant support in this journey.

Editor

Introductory Chapter

Overview

Konstantin Volkov

School of Mechanical and Automotive Engineering,
Faculty of Science, Engineering and Computing, Kingston University, London
UK

1. Introduction

Gas turbines are a vital and active area of research because they play a dominant role in the fields of power, propulsion and energy. They are used from the simple cycle machines employed to compress gas, pump oil and provide power, to the combined heat and power gas turbines used to provide electrical power, heating and cooling for industrial plants. Gas turbines are widely used in power plants and mechanical drive applications and, as these plants can be configured in a number of ways, the gas turbine manufacturer needs to balance the requirements of each user to optimize the design.

The conceptual design process of gas turbines is complex, involving multiple engineering disciplines. Aerodynamics, thermodynamics, heat transfer, materials science, component design, and structural analysis are a few of the fields employed when down selecting an appropriate gas turbine configuration. Because of the complexity involved, it is critical to have a process that narrows gas turbine options without missing the optimum.

The robustness of a design process is dependent on a number of factors including clear requirements and objectives, capture of the design parameters, knowledge capture and dissemination, validated procedures, repeatability, manufacturability, and the capability to consider the widest possible scope in the search for a conceptual design solution. The use of a constraint modelling technique has provided a framework where the various elements and tools involved in a design process can be integrated through various communication methods.

The design parameters of the gas turbines need to be chosen carefully to balance their influence on the reliability, maintainability, cost, efficiency and emissions of a gas turbine based power plant. Efficiency and reliability are two major parameters that should both be considered at the beginning of a new design project. To get higher efficiency means higher firing temperatures, higher pressure ratios, exotic materials, complicated cooling systems, all factors which jeopardize the cost and the reliability of the product. The aim of the product design team is to reach the optimum balance for these parameters, and for the demands and specifications of the individual customer. The gas turbine design process is not completely linear since the design steps are highly interdependent. A number of iterations are usually necessary in selecting a final configuration.

This book focuses on development and improvement of methods and techniques of analysis and diagnostics of efficiency, operation and maintenance of gas turbines. Authors from several countries have contributed chapters dealing with a wide range of issues related to analysis of gas turbines and their engineering applications. Gas turbine engine defect diagnostic and condition monitoring systems, operating conditions of open gas turbines, reduction of jet mixing noise, recovery of exhaust heat from gas turbines, appropriate materials and coatings, ultra micro gas turbines and applications of gas turbines are discussed. Analytical and experimental methods employed to identify failures and quantify operating conditions and efficiency of gas turbines that are encountered in engineering applications.

The book contains 11 chapters written by the specialists from various countries who are working in field of design, optimization, maintenance and diagnostics of gas turbines.

2. Ultra-micro-gas-turbines

Ultra-micro gas turbine generator, that is a power device with high power density, is analysed in this chapter. This generator, although the covered power range oscillates between 100 and 500 W, is characterized by reduced overall dimensions. Design issues and realization of the mechanical components is considered. The economic impact of these devices depends on the performance levels and the manufacturing costs, both of which have yet to be proven. Competitiveness of ultra-micro gas turbine generator with conventional machines in a cost per installed kilowatt is discussed.

3. The selection of materials for marine gas turbine engines

This chapter presents hot corrosion results of selected nickel based superalloys for marine gas turbine engines both at high and low temperatures. The results are compared with a new alloy under similar conditions in order to understand the characteristics of the selected superalloys. It is observed that the nature and concentration of alloying elements mainly decide the resistance to type I and type II hot corrosion. Relevant reaction mechanisms that are responsible for degradation of various superalloys under marine environmental conditions are discussed. The necessity to apply smart coatings for their protection under high temperature conditions is stressed for the enhanced efficiency as the marine gas turbine engines experience type I and type II hot corrosion during service. Hot corrosion problems experienced by titanium alloy components under marine environmental conditions are explained along with relevant degradation mechanisms and recommended a developed smart coating for their effective protection.

Two chapters were written by the same authors and focus on energy and exergy analysis of the Brayton cycle and operation of gas turbines in hot humid and arid climates.

4. Energy, exergy and thermoeconomics analysis of water chiller cooler for gas turbines intake air cooling

Gas turbine power plants operating in arid climates are considered in this chapter. They suffer a decrease in output power during the hot summer months because of the high specific volume of air drawn by the compressor. Energy and exergy analysis of a Brayton cycle coupled to a refrigeration air cooling unit is discussed and shows a promise for increasing the output power with a little decrease in thermal efficiency. A thermo-economics algorithm is developed to estimate the economic feasibility of the cooling system. The cost of adding the air cooling system is also investigated and a cost function is derived that incorporates time-dependent meteorological data, operation characteristics of the gas turbine and the air intake cooling system and other relevant parameters such as interest rate, lifetime, and operation and maintenance costs.

5. Energy and exergy analysis of reverse Brayton refrigerator for gas turbine power boosting

The use of reverse Brayton cycle to boost up the power of gas turbine power plants operating in hot humid ambiance is discussed in this chapter. The effects of irreversibilities in the system components (air compressor, combustion chamber, turbine, air cooler, expander and the mixing chamber) are evaluated along with the exergetic power gain ratio and the exergetic thermal efficiency change of the cycle. The dependency of the power gain, thermal efficiency and exergetic efficiency on the operation parameters are presented and analyzed.

6. Gas turbines in unconventional applications

Unconventional gas turbine applications are discussed in this chapter. Some of engineering solutions are intended for smaller gas turbine systems, where the regenerative heat exchanger supplies energy for an additional thermal cycle. Coupling of Brayton cycle with several other thermodynamic cycles (e.g. another Brayton, Diesel or Stirling cycles) is discussed, and advantages of hybrid systems are analyzed. Large international development programmes are reviewed, and several hydrogen-fuelled gas turbine concepts are proposed. Potential combination of a hydrogen-fuelled gas turbine and a nuclear power generation unit which is used to cover peak load power demands in a power system is described.

7. The recovery of exhaust heat from gas turbines

In this chapter different techniques for recovering the exhaust heat from gas turbines are discussed, evaluating the influence of the main operating parameters on plant performance. A unified approach for the analysis of different exhaust heat recovery techniques is proposed. The methodology is based on relationships of general validity and characteristic plane for exhaust heat recovery, that indicates directly the performance obtainable with different recovery techniques, compared to a baseline non-recovery plant. An innovative scheme for external heat recovery is presented. It envisages repowering existing combined cycle power plants through injection of steam produced by an additional unit consisting of a gas turbine and a heat recovery steam generator.

8. Gas turbine diagnostics

This chapter focuses on reliability of gas path diagnosis. New solutions are proposed to reduce the gap between simulated diagnostic process and real engine maintenance conditions. Thermodynamic models, data validation and tracking the deviations, fault classification, fault recognition techniques, multi-point diagnosis, diagnosis under transient conditions, and system identification techniques are presented. Practical recommendations are given to develop an effective condition monitoring system.

9. Models for training on a gas turbine power plant

In this chapter a summary of the gas turbine simulator development and its model characteristics are described. Stochastic and discrete events models are not considered, but deterministic models of industrial processes are contemplated. The simulator was tested in all the operation range from cold start to 100% of load and fulfils the performance specified by the client, including a comparison of its results with plant data.

10. Summary

Many different methods exist to integrate various design elements into an overall process. Ideally, designers like to perform all design steps concurrently in order to minimize the overall time required to conduct a study. Whole books have been written to address each of steps involved in design of gas turbines, so this book cannot possible cover all issues. Hopefully, it has at least introduced the reader to tools that are currently available.

The book covers many aspects of gas turbine design and operation. The book represents the latest research of various groups of internationally recognized experts in gas turbine studies. This book is intended for engineers and technical workers in design, optimisation and maintenance of gas turbines, and specialists is thermodynamics and heat transfer, particularly those involved with energy systems and transportation systems that make use of gas turbines. It will be of interest to academics working in aeroengine control and to industrial practitioners in companies concerned with design of gas turbines. The works presented in the book are easily extendible to be relevant in other area in which gas turbines play a role such as power engineering and marine engineering.

The open exchange of scientific results and ideas will hopefully lead to improved reliability of gas turbines and aeroengines. The book presents necessary data and helpful suggestions to assist scientists and engineers involved in the design, selection and operation of gas turbines.

Ultra Micro Gas Turbines

Roberto Capata

Department of Mechanical and Aerospace Engineering, University of Roma 1,
Faculty of Engineering, Roma
Italy

1. Introduction

1.1 State of art

Object of the present work is the detailed study, in every its aspect, of Ultra-Micro-Gas-Turbine Generator, that is a power device with high power density. These generators, although the covered power range oscillates between 100 and 500W, is characterized by very reduced overall dimensions: this introduces complications in the design and, above all, the realization of the mechanical components who represents the greater difficulty to exceed. The advanced searches in this field preview the realization of the characteristic structures of the machine with high tech systems:

- the impellers, both turbine and compressor, can be manufacturing with silicon, titanium or special alloys micro laser techniques (figure 1);
- the combustion chambers can be obtained in toroidal spaces with the possible lowest volume: this involves the necessity to construct fuel injection device technologically complicated, inside the same combustion chambers; a different model is constituted, in some NATO devices, by a 2-D geometry combustion chambers (an example is visible in figure 2);
- the electrical generators are constituted by a highest number of polar braces due to high rotational speed of the machine (> 100 000 rpm). Their realization is realized by micro manufacturing of planned rings that have been inserted in the case (for the statoric part), and the other ring is located on the faces the rotor impeller (usually the compressor, to avoid the high temperatures problems);
- the rotational speed is within 150 000 ÷ 350 000 rad/s (in some cases > 500.000)

a)   b)

Fig. 1. Typical compressor a) and b) turbine D geometry [Epstein 2003]

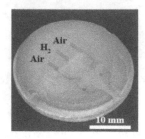

Fig. 2. Typical 2D combustor

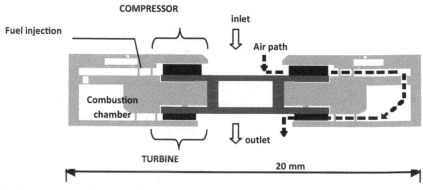

Fig. 3. MIT prototype [Epstein 1999 & 2003]

In figure 3 it shown a prototype realized from the M.I.T., that can be consider as the "symbol/emblem" of these researches carried out during the last few years in micro-turbines. It is constituted by a overlapping layers in sequence, starting from the compressor, the toroidal combustion chambers and the turbine. The characteristic data of this prototype can be reassumed in:

- Fuel: Hydrogen
- Fuel consumption: 16 g/h
- Power: 4 ÷ 10W
- Rotational speed: ≈10^6 rpm
- TIT: 1600 K
- Used materials: Si And SiC

We can notice the temperatures reached in these systems are higher than those characteristics of the actual large scale systems. Therefore, in the zones affected by the highest thermal stresses, the Silicon Si has been replaced with the most resistant SiC material. This prototype represents the highest state of the technique and the more advanced research that can be reached in this moment. Other examples of UMGT exist, however, thus do not characterized by exasperated technological levels and with a similar structure to the conventional systems. Also regarding the used materials, these device use typical steels of the commercial impellers. The thermodynamic parameters are similar to large scale sets:

- Compression ratio: ≤ 3

- Air mass flow rate: ≈ 2 g/s
- Rotational speed: $\leq 300\ 000$ rpm
- TIT: ≈ 1300K
- Net power: $4 \div 100$ W

2. Scaling factor

Since 90's, the academic world started to study the possibility to realize a "micro" gas turbine (GT) set, with an overall dimensions of the order of the centimetres, able to developing a nominal power within 0.10 to 100 kW. Such a type of apparatus are defined today "micro gas turbine". At the moment, the larger GT sets in service are characterized by a radial dimensions of the order of the meter and develop hundred of MW of nominal power, while the generator of a UMGT is large few millimetres, for which the elaborated mass flow rate by the micro device will be 10^{-6} times lower than a conventional machine, if the same tip speed is maintained. Moreover, the generated power will be 10^{-6} lower, consequently a micro GT device would be able, in first analysis, to producing about hundred of Watts. This value re-enters in the extremely extended field of the portable equipment applications in which the battery packages represent the main solution to the problem. The obtainable power density is greater than a conventional battery package, thanks to the fact that the power scales with the flow rate and ,therefore, with the square of the length, while the volume with the cube:

$$P \propto L^2 \tag{1}$$

$$V \propto L^3 \tag{2}$$

so:

$$P / V \propto 1 / L \tag{3}$$

Conceptually it is possible to realize any thermodynamic cycle to these scales, but it is necessary to remind that the reduced dimensions do not allow the possibility to insert any components characterized by particularly complex geometries. This consideration forces the research to adopt a simple Brayton cycle. Other considerations concern the fluid dynamical behaviour of the device and the mechanics of the machine, which will modify the relative choices to the thermodynamic cycle, moving the optimal design parameters in different zones from those developed for a classic large scale machine. In fact, while the speed and the temperatures are almost unchanged regarding the macro turbines, the chord blade is of the order of millimetre. The consequence is that the value of Re number is lower than the actual conventional large scale turbo machines (usually between 10^5 - 10^6). Viscous forces will have greater influence and the relative losses will be more evident. The losses linked to the three-dimensional aspect of the device, to the clearance between case and disc or to shock waves will be more significant and higher than the common ones. It is reasonable to think that the increase of the viscous forces produces higher losses for aerodynamic resistance in the piping and/or manifolds. One of the problems that the research/designer must be faced in the study of the micro fluid dynamic phenomena is the adequacy of the continuous fluid model. To evaluate the possibility of adopting the Navier-Stokes equations, the Knudsen number Kn, equal to the ratio between the free medium path of particles and characteristic length, can be used. Several tests have evidence that the first effects of the non-continuity appear for values of Kn = 0,1, while the continuous model loses its validity to Kn = 0.3. Considering that the free medium path of the air, at atmospheric conditions, is 70 nm (nanometres), the flow will have to be considered discontinuous for lengths of 0.2 μm. The considered micro devices dimensions (the gap will be of the order of some μm) the flow can be still studied as continuous, therefore it is not necessary to consider the molecular kinetic. The heat exchange between the device components and with the fluid is higher, due to the small scales of length, for which the thermal gradient will be lower and consequently thermo-mechanical stress will be lower. On the other hand, the parts isolation will be more complex and the heat losses towards the outside will be higher. The materials deserve a detail attention since the reduced scales allow the introduction of light ceramic material, with very attractive mechanical properties, but - to large-scale -unusable for the GT construction. With these dimensions, in fact, the material can be still considered as a continuous one, for which the plastic, the elastic, the creep and the oxidation behaviour, and finally the coefficient of thermal conduction, do not change, while the mechanical resistance is greater, strongly dependent on the manufacturing defects that are "naturally" limited by the micro-dimensions.

3. Thermodynamic cycle

Generally, the thermodynamic analysis of a system is independent by the scaling factors; theoretically it is possible to adopt any type of cycle to the micro-scales without particular attention. The technological limits, the inherent mechanical and fluid dynamical considerations and the necessity of device compactness, impose limited and important choices. In the case of the UMGT, it has been decided to privilege, fundamentally, the compactness and simple manufacturing, to contain the costs, admitting not optimal machine efficiency. This choice does not involve any particular disadvantages, since from theoretical extrapolations show sufficient efficiencies, to obtain higher energy density than to the best actual battery packages. The thermodynamic cycle adopted, in terms of simplicity, is the Brayton cycle, that has moreover the advantage to supply increasing energy density with the increase of the operational speed. The main disadvantage is the necessity in such case to have a component efficiency at least to the 40-50% for being self-sustainable: it must, therefore, be headed at most efficient and possible productive technologies. The state of the art for thus small structures renders the integration of conventional cooled or regenerated cycles improbable, but remains the possibility to use the outlet fluid of the compressor to cool the walls of the combustion chamber, or to obtain a sort of pre-heating, without to complicate geometry. To obtain a thermodynamic cycle adapt to the prefixed scope, the outlet temperatures of the combustion chamber must be comprised between 1200-1600 K and the tip speed between 300-600 m/s, consequently stresses will be of the 10^2 order of MPa for compression ratios between 2:1 and 4:1. Wanting, at least, to privilege compactness and lightness the choice of radial machine is most suitable, thanks the possibility to use a single stage for compressor and turbine. The influence of the Reynolds number on the performance of turbo machines has received considerable attention in literature, and several more or less general and sufficiently reliable models have been demonstrated, both for incompressible and compressible flows. Historically, the first formulations originated from empirical data collected on hydraulic turbine models and yielded acceptable results for full-scale applications. It is clear that the scaling for other types of machinery -and specifically for radial pumps and compressors- was much more sensitive to additional factors, e.g. the different relative roughness of the full-scale machine with respect to the model, leakage effects due to the geometric non-similarity of machining tolerances and clearances, and both the lack of data and the excessive measurements uncertainty in the smallest models. There have been numerous attempts to formulate a general model of $\eta = f$ (Re) scaling, but all the proposed models agree with the experimental data only within a limited range of configurations and fall therefore short of representing general design correlations. The main problem of the available formulations seems to be on the one side their complexity and on the other side the non-crisp phenomenological model they subsume, but the common point in all approaches is the distinction in Reynolds dependent and independent efficiency losses. This assumption leads to unavoidable difficulties in the determination of the Re-independent loss fraction, mostly originated by non-homogeneous factors like manufacturing methods, tolerances and clearances. An interesting series of experimentally validated studies were conducted in the 80's for single and multistage centrifugal compressors by Wiesner [Wiesner 1979], Casey and Strub: the results of these investigations differ from Author to Author, as do their respective conclusions, but one common point all agreed upon is the definition of a reference Reynolds number based on the width of the exit section:

$$Re = \frac{U_2 b_2}{\nu} \tag{4}$$

By contrast, since we are interested here in introducing correction factors to the Balje charts, it was necessary to adopt the same choice of parameters: therefore, in this work the *Re* definition as reported in [Balje 1981] has been used throughout:

$$Re = \frac{U_2 D_2}{\nu} \tag{5}$$

Some scaling formulations include the influence of Reynolds number and surface roughness, but in this work we consider only hydraulically smooth surfaces, eliminating thus the roughness variable from the picture, also in consideration of the rather scant data available on the few prototypal rotors in the ultra-micro scale range. The derivation of a "universal" formula is intrinsically difficult, due to the substantial difference in the flow phenomenology, so that each class of machines requires a specific analysis. The objective of this work is to propose a preliminary method to scale the efficiency of ultra-micro-compressors and turbines, investigating the possibility to extend the applicability of Balje charts for miniaturized machines, to reduce design time- and resources investment.

3.1 Reynolds number effects – Problem formulation and coefficient definition

The general functional relationship proposed in the available references is of the so-called Stodola form:

$$\frac{1-\eta}{1-\eta_{ref}} = a + (1-a)\left[\frac{Re_{ref}}{Re}\right]^n \tag{6}$$

Whereas the coefficients a and n differ from Author to Author, as represented in Table 1.

Year	Source	Inviscid loss fraction"a"	Viscosity-dependent loss fraction (1-a)	Exponent n	Machine type
1925	Moody	0.25	0.75	0.33	Propeller turbines
1930	Ackeret & Muhlemann	0.50	0.50	0.20	Hydraulic turbines
1942	Moody	0.00	1.00	0.20	Pumps
1947	Pfleiderer	0.00	1.00	0.10	Pumps
1951	Davis, Kottas & Moody	0.00	1.00	variable	All turbomachines
1954	Hutton	0.30	0.70	0.20	Kaplan turbines
1958	Rotzoll	0.00	1.00	variable	Pumps
1960	Wiesner	0.50	0.50	0.10	Radial compressors
	Fauconnet	0.24	0.76	0.20	
1961	O'Neil & Wickli	0.00	1.00	variable	Radial compressors
1965	ASME Code PTC-10	0.00	1.00	0.20	Axial compressors
		0.00	1.00	0.10	Radial compressors
1971	Mashimo et Al.	0.25 min	0.75 max	0.20	Radial compressors
1974	Mashimo et Al.	0.15-0.57	0.43-0.85	0.20-0.50	Radial compressors
2007	Capata, Sciubba & Silva	0.50	0.50	0.25	All turbomachines

Table 1. Summary of the most popular efficiency correction equations

Where the coefficient a represents the Reynolds number independent loss fraction, but it is in fact sometimes itself a function of Re, and the exponent n, as proposed in Strub [Strub 1987], is in general taken to be inversely proportional to the Reynolds number: this approach accounts for the decreasing influence of the viscous losses for high values of Re. The main shortcomings of this approach are:

- The coefficient a is likely to remain constant only in a small range of Reynolds numbers;
- The Re-independent losses are also related to other factors like for example leakage;
- It is also well known that a varies with both geometry and manufacturing techniques even for machines with the same Re. As a first approximation, we assumed that the manufacturing process and clearances fit with the parameters imposed for the Balje charts. This assumption is not far from reality, because the design of such small devices is strongly constrained by factors unrelated with fluid-dynamics, so that ultra-micro-machines are indeed all quite similar to each other, and because the estimated clearances [Epstein 2003] are in the same range as those reported on the Balje charts.
- The exponent n varies considerably from author to author and this might depend on the different intrinsic accuracy of the data used by different authors, originated by the neglection of the effects of relative roughness. As stated above, in this work we concentrate on the influence of the Reynolds number, thus as a first step we, too, neglected the relative roughness, especially because we could find no experimental data in the range of the geometrical scales of importance here. Wiesner [Turton 1984] pointed out that it is useful to introduce an explicit Re-dependence in this exponent, and proposed a functional relation of the form:

$$n = k' \left(\frac{1}{Re/Re_{ref}} \right)^k \tag{7}$$

Notice that Equations (6) and (7) appear to be "complex" enough (i.e., to subsume a deep enough phenomenological model) to yield a good estimate of the losses for a particular class of machines such as the ultra-micro ones considered here. To circumvent the lack of a large experimental database, we decided to follow a different "type" of empirical experiments, in which numerical simulations substitute for laboratory tests. A sufficiently large set of data was created by means of accurate numerical simulations, and values for the constants a, k and k' in equations 6 and 7 above were obtained by a best fit procedure. Thus [Capata & Sciubba 2007] the final formula we propose for the scale-down is:

$$\frac{1-\eta}{1-\eta_{ref}} = 0.50 + 0.50 \cdot \left[\frac{Re_{ref}}{Re} \right]^{\left[0.084 \cdot \left(\frac{Re_{ref}}{Re} \right)^{0.25} \right]} \tag{8}$$

The coefficients in equation (8) provide a good agreement with the values reported in literature, particularly with the model by Wiesner. An direct confirmation of the validity of our procedure was obtained by applying equation (8) to compute the efficiency of an ultra-micro compressor designed by the research group at MIT. Finally we suggest to adopt a polytropic efficiency within 0.55 and 0.7 in the preliminary design of the turbomachines.

3.2 The thermodynamic CYCLE – General overview

The cycle equation sets are:

$$p_2 = p_1 \beta_c \tag{9}$$

$$T_2 = T_1 \beta_c^{\frac{k_c-1}{k_c \eta_p}} \tag{10}$$

$$T_4 = T_3 \left(\frac{1}{\beta_t} \right)^{\frac{(k_t-1)}{k_t} \eta_p} \tag{11}$$

where p_1 is the inlet static pressure equal to 1 atm, p_2 is the compressor outlet static pressure, T_1 and T_2 the temperatures of beginning and end compression process respectively, and T_3, T_4 refer to the turbine expansion. The compression ratio $\beta_c = 2$ is different from the expansion ratio $\beta_t = 1.94$, due to the combustor losses (approximately 3%). Once that the static temperatures in the main points of the cycle are known, it is possible to evaluate the mixture ratio α and the equivalent ratioϕ. Then, the compression work W_c the expansion one W_t and combustion process Q, can be calculated:

$$W_c = \int_{T_1}^{T_2} c_{P_{air}} dT \tag{12}$$

$$W_t = \int_{T_3}^{T_4} c_{P_{mix}} dT \tag{13}$$

$$Q = \int_{T_2}^{T_3} c_{P_{mix}} dT \tag{14}$$

remembering that the molar fractions (necessary to calculate the $c_{P_{mix}}$) are obtained from the equivalence ratio:

$$\frac{n_{C_nH_m}}{n_{air}} = \phi \left(\frac{n_{C_nH_m}}{n_{air}} \right)_{stech}$$ (15)

$$x_{C_nH_n} = \frac{n_{C_nH_m}}{n_{C_nH_m} + n_{air}}$$ (16)

$$x_{air} = 1 - x_{C_nH_m}$$ (17)

Once defined the net work W_u, it is possible to determine the actual thermodynamic efficiency η_{th} :

$$\eta_{th} = \frac{W_u}{Q}$$ (18)

The net power can be calculated as follows:

$$P = \eta_e \dot{m}_{air} \left[\left(1 + \frac{1}{\alpha} \right) \eta_m L_t - \frac{L_c}{\eta_m} \right]$$ (19)

$$\dot{m}_{C_nH_m} = \frac{\dot{m}_{air}}{\alpha}$$ (20)

And the global efficiency η_g is:

$$\eta_g = \frac{P}{LHV \cdot \dot{m}_{fuel}}$$ (21)

4. UMGT component analysis

The first goal in the design of an UMGT is to contain the device overall dimensions and to avoid, if possible, a multi stage configuration: wanting to obtain a reasonable compression ratios, the choice falls on radial machine (centrifugal compressor and centripetal turbine). Being the thermodynamics invariant regarding the macro turbines, to generate the required specific power with single stage machines, it will be, however, necessary to adopt compression ratios between 2:1 and 4:1 and TIT within 1200 - 1600 K. Higher temperatures has to be avoided, due to the difficulty to integrate a cooling system. The centrifugal stresses (most important), as the specific work, depends on the square of the speed, that implies for, the considered values of U (velocity), between 300 and 600 m/s, stresses in the order of the hundreds of MPa. Not being able to take advantage of the conventional turbomachines manufacturing techniques and considering the various fluid dynamic problematic at the micro scales, it will be necessary to study which parameters of the conventional design are still adaptable and, where is necessary to act ex-novo in order to redefine an optimal area of design. The most important constrain in the design procedure is the adoption of 2-D extruded geometries for the blades, using the photolithography techniques, now widely diffuse for the production of MEMS (Micro Electro-Mechanical-Systems). Fortunately continues steps in the development of such manufacturing techniques has been recognized and more complex 3-D geometries can be, nowadays, produced. An other design constraints are the maximum obtainable blades heights. This value, with the laser techniques, was limited to 500 μm few years ago, but now it is possible to obtain higher values, clearly, with greater costs. The Reynolds number is lower than conventional machines and considering the several constructive limits, will be difficult to succeed to create a geometry to avoiding separation, moreover the net output power will be probably lower than the hypothetical one with simple scale considerations. The work will be, substantially, produced by the mass forces. The conventional radial machines, usually, are equipped with inducer or exducer and, guided through sweet bending of the channels through the rotor, in such way to guarantee the lower possible losses for the interaction between boundary layer and angles or the flow separation: in this case it could not be constructively simple or favourable economically to realize a rounded inlet manifold or an axial outlet channel. Considering the diffusive process of the compressor, it could be useful to increase the chord length and to compensate the absence of constriction in the channel, increasing the blade thickness, in particular the trailing edge will be thicker than conventional machine. Consequently, the more opportune choice could be the reduction of the gap between the rotor and the case, to decrease the ventilation losses, but to minimize such distance implies to increase the resistance losses for the low blade height. Recent studies of the MIT have evidenced that a good strategy to such scale is to realize a gap equal approximately to 1% of the blade height. The problem of the flow separation in turbine is less problematic than in the compressor. The Reynolds number value is about 10^4 or lower, due to the elevated temperatures produced by the combustion. The significant losses will be at the flow exit, due to the straight angle (90°) and of the swirl residual. To partially recover this residual a diffuser will be insert.

4.1 Combustion chamber at MIT

Undoubtedly, the combustion chamber plays a crucial role in the design of the entire device, considering, above all, the necessity to limit the dimensions. The main requirements of a combustor are high efficiency, low pressure drops, high structural resistance and low emissions. Surely, for a conventional combustor is simpler to obtain these requirements, thanks to the ability to realize - to the macro scales - complex structures. In favour of a micro combustor there is an higher energy density, coupled to a greater elaborated flow rate for volume unit, as can be noticed in table 2.

In a UMGT, the relative volume, will be, approximately, 40 times greater than a traditional system. The largeness of the combustion chamber, in fact, is dictated by the necessity to completely develop the reactions and by the residence time, that it is the sum of the necessary time to the mixing (that scales with the dimensions of the device) and the chemical reaction time (that is fixed and forces to consider greater volumes at the micro scales). In a conventional engine, beyond 90% of the residence time is reserved to the mixing. At this point it is useful to introduce the number of Damkohler "Da", that it is the ratio between the fluid dynamic time, (how much time the fluid spends to cross the combustion chamber) and the chemical reaction time. Experimentally it is verified that, for having a complete reaction, it is necessary to have Da number higher the one (Da > 1). In a combustor is fundamental to obtain the highest possible energy density, and to achieve this objective, is necessary to be capable to maintain high flow rate for volume unit. Moreover is necessary to avoid to increase the chamber dimensions. At the same time it is fundamental

	Conventional Combustor	Micro combustor
Length	0.2 m	0.001 m
Volume	0.073 m³	6.6×10^{-8} m³
Cross-sectional area	0.36 m²	6×10^{-5} m²
Inlet total pressure	37.5 atm	4 atm
Inlet total temperature	870 K	500 K
Mass flow rate	140 kg/s	1.8×10^{-4} kg/s
Residence time	7 ms	5 ms
Efficiency	> 99%	> 0.9
Pressure ratio	> 0.95	> 0.95
Exit temperature	1800 K	1600 K
Power density	1960 MW/m³	3000 MW/m³

(note: the residence times are calculated using inlet pressure and an average flow temperature of 1000 K)

Table 2. Comparison between a conventional combustor and a micro combustor [MIT prototype – Epstein 2003]

to maintain the Damkohler number higher than the unit. It can act in two ways: increase the fluid dynamic time to complete the combustion or decrease the necessary residence times. Clearly, the first choice contrasts with requirement of maximum energy density, since previews the greater dimensions, while, the second one, is applicable with good result by the use of catalysts. The greater surface/volume ratio of a micro combustor increases the efficiency of catalysts, accelerating the chemical reactions. In this case the heat losses are not negligible, as well as in the traditional systems, since an overall efficiency and reaction temperature reduction have been produced. In effects it has been demonstrated that the relationship between lost heat and generated heat is inversely proportional to the inverse of the hydraulic diameter of the combustion chamber:

$$\frac{Q_{lost}}{Q_{prod}} \propto \frac{1}{D_{hyd}^{1.2}} \tag{22}$$

The hydraulic diameter of a micro combustor is of the order of millimetre, one hundred times lower than conventional one. This fact implies that the heat losses correlated to the heat produced by the combustion will be approximately one hundred times higher than the traditional combustors. For these reasons will be improbable to achieve an efficiency of 99% compared to the commercial systems. Finally, a reaction temperature drop increases the reaction time. All these factors reduce the available "design space", as consequence of an important decrease of the Damkohler number, and the flammability range, due to the reaction time decrease. Clearly the choice of fuel completely modifies the design limits and, maybe, the conditions will be more restrictive. An effective way to reduce the residence times is the insertion of a premixing device, for example immediately after the compressor exit, after the "flame holder", so that the mixture reaches the combustor very stirred: this can produce backfire problems and instability, but the residence time is reduced of a factor 10 regarding conventional 5-10 ms. A method to reduce the heat losses and, at the same time, to cool the combustor walls consists in leading the compressor outlet air around to the external wall of the chamber. This technique derived directly from the conventional inverse flow combustor, the only difference is that the liner is cooled for conduction rather than through a fluid film. The configuration with premixing device previews the fuel injection before the chamber inlet, to begin the mixing with the air and to reduce the dimensions of the device. The relative technologies to the photolithography are extremely favourable for the injectors manufacturing. The choice of the materials represents an other design crucial point; the silicon introduces serious problems of plastic deformation at 950 K. A combustor with a dilution zone (for cooling the flow) can be adopted, and materials like silicon carbide SiC presents an higher thermal resistances. Remember that the thermal fluxes, previously discussed, facilitate the walls cooling. The dilution ("Dual zones") extends the range of the elaborated flow rate, but at the same time decreases the obtainable maximum efficiency of 10-20%. These combustors have been tested with an outlet temperatures of 1800 K approximately.

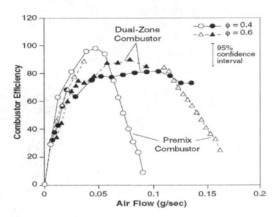

Fig. 4. Combustion efficiency of an hydrogen micro combustor [Peirs 2003, Epstein 1997, 2003]

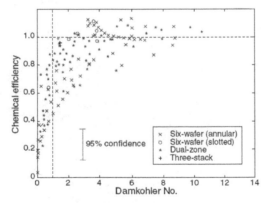

Fig. 5. Combustion efficiency to varying the Damkohler number

The necessity to assure the flame stability introduces more difficulties, so the configuration with dilution zone facilitates the recirculation in the primary zone through the air that enters in the secondary zone. Moreover the mixture ratio (in this case close to the steichiometric one) allows to decrease the reaction times. The adoption of the slots increases the turbulent phenomena in the chamber. Other fuel (as ethylene, propane, butane, methane, ecc) and other configurations (with platinum catalysts) have been tested and all the data are collected in Figure 5 figure according to the Damkohler number. It is evidenced how it is necessary to work with Damkohler numbers at least equal to 2, to obtain good combustion efficiency.

4.2 Other combustion chamber

Here follow is reported a particular case of a built and tested ultra micro combustion chamber.

4.2.1 UDR1 UMGT combustion chamber

The 2.5 kW UDR1 UMTG is composed of a compressor, a compact air pre-heater (the pressurized air exiting from the compressor is heated by ΔT= 393 K), a combustion chamber and a turbine, that together compose the thermal section; a double effect (reversible) electrical engine for the electric part; a shaft and the bearings for the mechanical, and obviously all the auxiliaries components required, as valves, controllers, and ducts, including the management unit. For safety reasons the cylindrical fuel tanks are external to the metallic UMTG enclosure [Capata & Sciubba, 2010].

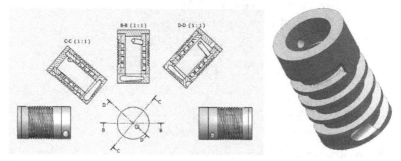

Fig. 6. CAD rendering of the combustor with a detailed view of the pre-heater.

The overall length of the assembled and modified combustion chamber is 2.2 cm, the outer diameter 4.2 cm and the overall height is 13.0 cm.

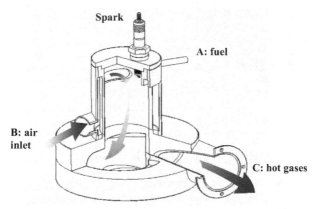

Fig. 7. The tested prototype of the combustion chamber

4.3 Bearings and rotor dynamics

A UMGT is characterized by highest angular velocities (1 or 2 orders of magnitude higher than conventional ones) and by clearance between the components of the order of the micron. It is therefore from is easy to verify as the great part of the design development is dedicated to the bearings and to the rotor dynamics. The first reflection is, due to the high rotational speed (> 20000 rad/s), the device will work in supercritical conditions, with operational frequencies higher than the resonance ones and for this reason the study of the stability will be primary in the machine design. Clearly the bearings play a dominant role in the rotor dynamics as they must support axial and radial thrust of the rotor, dampen the oscillations and support the forces deriving by the accelerations of the rotary parts, without to count the electric forces and the operational pressures. It cannot certainly be neglected that the supports must be able to supporting the high turbine temperatures and the thermal gradient presents and, then, be able to work in any operational condition. At last the resolution of all these problems must be consider the actual technological limitations. Other devices MEMS with rotary parts currently exist, but the angular velocities are such to allow to using not-lubricated bearings. Moreover do not exist in commerce such "little bearings" to be used for these rotors. Honestly, we have to mentioned two manufacturers that guarantee dry bearings at 200000 rpm (20933 rad/s). Therefore, the academic study focuses on two possible solutions: electromagnetic or air bearings. The first solution introduces the possibility to take advantage of magnetic fields or electrical ones to support the rotors loads. The problem relatively to the magnetic fields resides in the impossibility to adopt ferromagnetic materials in the productive technologies chosen (recording laser and photolithography), moreover the Curie point is such configuration not to allow of using these systems to the temperatures previewed in turbine and would be therefore necessary to introduce a cooling system, with all the design constraints and possibilities connected. Studies have been carried on the electric fields, but the forces that are succeeded to produce are decidedly inferior to support the considered loads. To all these consideration we must add that the electromagnetic bearings are strongly unstable and required a feedback control systems, which would complicate the installation. The air bearings introduce numerous advantages, in particular in terms of constructive simplicity, of loads capability, and relative insensibility to the problem of the high temperatures. Currently these device are already used for medium-small dimensions turbomachines. At smaller scales the gas bearings are wide used in micro gyroscopic system since many years. Wanting to make scale considerations it can be said that, to parity of load conditions, the ability of these bearings grows up with the overall dimensions decrease, because the ratio volume/surface decreases and consequently the inertial load. Certainly the rotor dynamics results to be, partially, simpler (the introduced structure is more rigid, if compared to the conventional ones), thus allowing the approximation of the rigid body. The development of the bearings has to preview two different use: first, the bearings must support the radial loads, second, must support the axial loads. These bearings are composed from a cylindrical hinge in contact with a lubricating support. More efficient models exist, but considering the complexity of the system, the more reasonable choice is the first one. The lubricating fluid in our case is air in pressure and according to how it is injected, the bearings are distinguished in "hydro-static" if the source is external and "hydrodynamic" if the support forces are directly generated by the spin of the disc (figure 8). Mixed configurations are possible (figure 9). Since the UMGT includes a compressor is possible to adopt both the configurations, and both solution are able to satisfying the loads requirements and temperatures stresses to these dimensions. The two types of bearings are characterized by various dynamic characteristics. In the hydrodynamic bearings the load ability grows up to increasing of the rotational speed, because the pressure of the film between disc and support increases with the spin. In theory, also for a hydrostatic bearing a mechanism of this type can be fed; we can imagine to support the gas in pressure through the compressor which will consequently regulate the pressure of the lubricating fluid. If the pressure of the fluid were maintained to constant the load ability would decrease gradually to increasing of the speed. In figure 10 is visible a comparison between the two types of bearings and a conventional one. In an hydrostatic bearings, pressure is maintained constant, while for an hydro dynamic one the load capability cargo depends on the speed and L/D ratio, where L is the length of the bearing while D is the diameter of the rotor. To decrease of this ratio the performances gets worse.

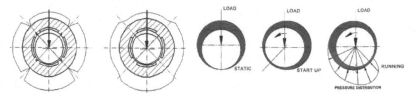

Fig. 8. Operational mode of a hydrostatic bearing (over) and hydrodynamic (under)

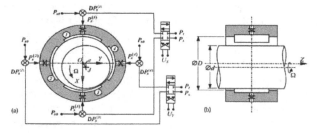

Fig. 9. Hybrid configuration

The fluid simply "sprayed" outside by the considered load, rather than to develop an adequate profile of pressures to balancing the external force. Unfortunately, in the laser manufacturing exist technological constraints, that does not allow to obtain elevated L/D ratio. Other parameters that influence the bearings design are the ratio between gap and the length of the bearing g/L, the mass of the rotor, the Reynolds number; while for the hydrodynamics bearings the load capability is dominated by the inverse of $(g/D)^5$ ratio. Currently the DRIE techniques allow to realize aspect ratio of 30:1, but a value of 100:1 can be reached. This means that assuming gap of 10-20 μm (that is the minimum value can be adopted to avoid the rotor failure for contact) can be had lengths of the channel of lubrication of 300-1000 μm. The design of an hydrodynamics bearing would suggest to us to decrease the g/D ratio and, therefore, to choose rotors with possible greater diameter, but this contrasts with the requirement of maximizing relationship L/D. Such parameters will probably find outside the normal design range. As already said the rotor operational frequencies are much higher than the resonance ones, and to such frequencies it is possible to model the bearing as a complex of springs and dampers whose effect is given by the lubricating gas which is cause also of operational instability due to transverse forces. The gap must be such to support a deviation of the rotor during the "crossing" of the critical frequencies without to hit the support. Figure 11 shows the MIT data (rotor diameter of 4 millimetre and gap of 12 μm). It can be noticed as the maximum displacement is lower than the maximum allowed. The data is in agreement with a famous analytical model, known as

"Jeffcott wheel": to low frequencies the rotor rotate around the geometric centre, while to at high frequencies rotates around the centre of mass. The dotted line represents just the loss of balance between geometric centre and centre of mass to which asymptotically stretches the rotor. For a hydrostatic bearing the critical frequency scale, simply, with the pressure and the viscous damping decreases to increasing of the speed. The amplitude of displacement of the rotor to the critical frequencies, then, increases with increasing of the pressure.

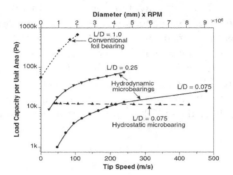

Fig. 10. Comparison between the load abilities of several bearings [Van den Braembussche 1993]

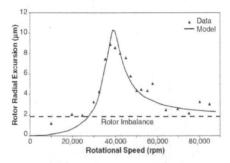

Fig. 11. Behaviour of an air micro bearing [Epstein 1997]

The trend will be similar to that reported in Figure 12, but to increasing of the pressure the peak will be moved up towards right and, this suggests to cross the critical frequency at low pressure values and low speed to, then, increase the pressure and to harden the bearing. A factor that influences the relative displacement amplitude at the critical frequency and the instability phenomena is the balance loss of the rotor. Normally the conventional machines adopt a dynamic balance that substantially consists in measuring the loss of balance and to modify the mass locally, to compensate the displacement between geometric centre and centre of mass. In the micro machines such problem is less sensible (the loss of balance for a rotor of 4 millimetre is about 1-5 µm) because the manufacturing process uses Si mono crystals or wafer of material much homogenous. To this can be add the manufacturing precision of ~1 µm. That it renders geometry much uniform. On the other hand must pay great attention to the alignment of the wafer. The hydrostatic bearings have the advantage of being stable to all the speeds within a fixed range, on the contrary the hydrodynamics ones suffer to low speed, while they are perfectly stable to the high ones. A way to stabilize these last bearings is to apply a unidirectional force that presses and pushes the rotor towards the wall of the support and, conventionally, the weight of the disc is used to generate such forces. The common method to quantify the decentralization necessary to create such forces is given by the eccentricity, ratio between the distance wall-rotor on medium gap (the 0 if the rotor is cantered, 1 if it touches the wall). The problem to the micro scales is that inertia is practically uninfluenced regarding the surface forces, this is favourable if we consider that the device demands orientation independence, but in terms of stability it has pushed to find such lay outs that take advantage of a heterogeneous distribution of the gas pressure to generate eccentricity. The zone behind the rotor is subdivided in two rooms that can be pressurized independently. Such system has been largely simulated and the result show that the rotor is stable for eccentricity values over 0.8-0.9. A "not yet solved" problem is related to the axial flows that pass in the interstice between rotor and support of the bearing: considering dimensions of the gap these flows do not seem negligible neither in hydrodynamic neither in that hydrostatic one. The stability, as we said, strongly depends on the $(g/D)^5$ ratio through a mass parameter:

$$\bar{M} = \frac{mp}{72L\mu^2}\left(\frac{g}{D}\right)^5 \tag{23}$$

It can be noticed that is necessary to have high eccentricities to having a gap, locally, more little. In this manner mass parameters decrease and, as consequence, the dotted line in figure 12 will be lower. The conventional turbomachinery succeed to obtain a minimal eccentricities closed to 0.5, but to our scales we are limited in the choice of the shape ratio.

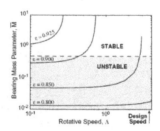

Fig. 12. Gas bearing stability in function of the eccentricity and normalized rotational speed [Freccette 2003]

Thus high eccentricities imply high accuracy in the circularity of the rotor parts and support, with the maximum shunting lines in the order of the μm, to avoid interferences. To adopt high eccentricities can be problematic in operating terms, because they produce a strong balancing losses. To obtain better performances could be the adoption of so-called "wave" bearings. This device is capable to dampening the whirl effects due to the excitation of the pressure perturbations generated by the particular geometry of the same bearings. This advantage does not involve some constructive complication and can be easy implemented through the photolithography techniques, however since the technological constraints define a minimal gap, the only way to create such geometry are to increase the interstices. From a performances analysis results that the load capability decreases, but the stability strongly increases and the minimal demanded load to stability operate (that means the minimal value to obtain the minimal stable eccentricity) decreases. Being the load capability of the bearings more sufficient than the required one, its decrease, in favour of the stability, would not be problematic for the correct operational mode of the device. Regarding the thrust bearings necessary to support axial stresses, both the hydrostatic configuration than the hydrodynamic one has been successfully tested and verified. In both cases the working point is sub critical, that is the exercise frequencies are under the critical ones, and this bring to an important design simplification. The hydrostatic bearings have a stationary behaviour and lead pressurized air in contact with the rotor through orifices circularly located inside the support. A value of about 2-5 atm (0.2÷0.5 MPa) are necessary to obtain adequate load capabilities and a good rigidity. The maximum rigidity is obtained when the pressure drop through the orifices equals the radial flow from the drainage edge of the bearing. The hydrodynamic bearings take advantage of the viscous resistance, exalted from a superficial envelope of spirals, to generate of the pressure gradient on the surface that increases towards the centre of the rotor. This mechanism self sustains and does not demand constructive complications, reducing the number of wafer necessary too. It is considered, moreover, that, while at the conventional scales, this type of bearings are less effective and not so used, at the small scales, thanks to the great increase of the superficial forces, regarding the volumetric ones, the load capability and the rigidity are similar to the ones of the high speeds hydrostatic bearings. This system introduces a problem of the minimal speed to generating a sufficient pressure to eliminate the rotor rubbing with the static parts: fortunately such minimal speed is of the order of 10^4 rpm [Epstein, Jacobson, Spakovszky] and this implies a malfunctioning limited to machine start up. The power dissipation of the two types of bearing has been calculated and verified by MIT institute and appears to be low. The necessity to obtain high accuracy in the bearings manufacturing, is in contrast with the demand of a device series production with an only one wafer. The plasma techniques, as well as all manufacturing techniques, introduce heterogeneous problems, that increase moving from the centre towards the periphery of the wafer. Regarding DRIE process, that is still in phase of maturation, seems that the manufacturing process carry to an inaccuracy in the recording depth of 3 µm on a diameter of 4 millimetre for the peripheral devices (the most penalized), than consist in a rotor loss of balance and, therefore, a total malfunctioning of the bearings. Actually MIT reports to have been tested this kind of bearings at a rotational speed of about 1.4 Mrpm [Epstein 2003]

4.4 Auxiliaries

The most important efforts in the UMGT design are dedicated, undoubtedly, to the turbomachines, the combustion chamber and the bearings. But we have to consider that the others components that completing the device will not be simpler designed or less important, being the system, in its entirety, a good compromise between the several parts, a compromise between the obtainable good characteristics from every single piece.

4.4.1 Electric generator and starter

The literature on the electric motors is extremely huge, but on the micro generators is insufficient. Seen the high temperature operational conditions, the necessity to integrate the components to avoid ulterior complications dictated by the bearings and the required power density, would be necessary an accurate study of the problem. Both magnetic and electrical motors are attractive as they are characterized by analogous power density, but the manufacturing difficulties correlated to the "not compatibility" of the ferromagnetic materials with the conventional micro technologies and the operational temperatures that reduce the properties of the same materials, have pushed to consider, in first analysis, the adoption of electrical devices. The power density scale with the square of the force of the electric field, with the frequency and the rotational speed. Numerous possible configurations for an electrical motor-generator exist; the first choice made by the MIT researchers is an induction machine, because this type of machine does not require a direct contact between the electrical device and the rotor neither the exact knowledge of the position of the rotor [Epstein 2003]. The rotor is composed by a layer of 5-20 µm of good insulator covered by a thin layer of a low conductor (high superficial resistance), while the stator is composed by a series of radial electrodes conductors supported by an insulating layer. The rotating electrical potential is imposed by the external electronics on the stator electrodes and the rotating electric field generates a distribution of charges on the rotor, which is mechanically driven. According to the relative phase between the motion of the charges on the rotor and the statoric field, adjustable from the outside, the device will operate as generator, motor or brake. The torque increases with the square of the force of the electric field and the frequency, but the maximum allowable force is regulated by the dimension of the gap. For the air the maximum is obtained for interstitial values of little micron, so that such machine can theoretically achieve an higher power density than the conventional device with analogous configuration. The frequency is regulated by the external electronic systems and by manufacturing device of the electrodes. Currently 300 volts and a frequency of 1-2 MHz is the maximum value obtained with a 6 mm rotor to producing 10 W with a gap of 3µm and a number of poles much high (beyond 100). To maximize the output power is necessary that the space between rotor and stator is of the same order of the statoric electrodes pitch, that means few µm, but with this type of gap the losses for viscous resistance, considering the machine rotational speed, are extremely high and represent the main source of losses. It is necessary to find a compromise between power density and efficiency. The viscous losses unfortunately represent more than the half of the total ones and limit the efficiency to 40-50%. A magnetic induction machine has the advantage of having less poles and optimal distances rotor-stator very wider (≈10 µm), that produce higher efficiency, around 60%. An ulterior advantage of these devices is that they operate at low frequencies and voltages, and are simpler to build. Currently the greater problems is the thermal resistance of the materials (actually 500 K, but material to working to 800 K have been currently studied) and the rotational speed in conflict with the necessity to insert on the rotor surface a iron layer of about 100 µm. Recently, thanks to new technologies of implantation of the ferromagnetic materials on silicon wafer, a good magnetic induction machine to with interesting efficiency and performances has been realized. As already said, to succeed to adopt a magnetic induction machine respect to analogous electrical devices, implies a greater efficiency, tolerances less limiting and higher manufacturing simplicity. An example of a magnetic induction machine is shown in the figure. The rotor diameter is 10 mm. The machine consists of a two phases stator and 8 poles and of a annular rotor [Epstein 2003]. The electromechanical conversion of the energy is achieved by the interaction of the magnetic field that evolve in the interstice rotor-stator with the current induced in the rotor the displacement of the magnetic wave. The stator consists of two phases composed by planar copper spirals, insert in three-dimensional blocks of vertically laminated ferromagnetic material, all supported by a silicon chassis. The ferromagnetic nucleus has an "onion" configuration, the sheets forms concentric rings, approximately 30 µm thick. This particular shape serves to reduce the induced current losses. The rotor is composed by 2 ferromagnetic ring with a thickness of 250 µm and 2 mm wide covered by a copper layer of 20 µm. The copper is extended over the external beam of the rotor for an ulterior millimetre, to exalt the induced current generation and to increase the maximum torque.

5. Materials

The problems connected to the choice of the materials for a UMGT are referred, partially, to those of conventional machine, in terms of mechanical stresses constraints, operational temperature and manufacturing processes. The materials, generally, satisfy some characteristics penalizing some others, therefore, it is indispensable in the preliminary design to find a good compromise, and not always, the better choice is univocal. It is considered, moreover, that according to manufacturing, the same material can introduce different characteristics. The properties of greater interest for the studied machine are the material absolute specific strength and its resistance to the thermal shocks at high temperatures, the creep and the oxidation, and its derangement characteristics under fatigue cycle. Considering the manufacturing processes and the device dimensions, the first choice for a device MEMS falls on the silicon, thanks the great maturity of the material productive technologies. The silicon also having of the good properties of specific strength and to thermal shock, presents a ductility and a strong "flauge" over the 550°C. This means that I can not use this material in the turbine blade manufacturing. On the contrary, the Silicon alloys appear more interesting, but the manufacturing of these materials is not still mature as well as the silicon ones. The metallic alloys are not able to operate at the demanded high temperatures without cooling or covering, that could create constructive complications. Moreover, the metallic manufacturing does not have a good accuracy. Other advanced ceramic materials, characterized by higher operational temperatures, have insufficient mechanical characteristics to the high temperatures. Adding the fragile behaviour of the ceramic materials, the manufacturing process results problematic and, at the moment, MEMS techniques are accurate only for Silicon and its alloy. Figure 13 put in evidence that the silicon alloys, in this case alumina, can be used at the high temperature, but the Al_2O_3 alloy has lower thermal conductivity and an higher thermal expansion. These characteristics renders the alloy particularly subject to thermal shocks and deformations. Also regarding hardness and elasticity modulus the SiC and Si_3N_4 have higher resistance performances, in particular to bending, and it can be noticed that, also having lower values to low temperatures respect to, for example, the zirconium, such value is maintained approximately constant to the high temperatures and is higher than the other materials. In figure 14 the SiC mechanical properties are maintained, substantially, unchanged with the temperature. The weak of the silicon alloy, like all ceramic materials, is the fragile behaviour, that to large-scale has delayed theirs uses. In large scale components, in fact, the inner imperfections can be in such amounts and largeness that is sufficient the effect of a small extra-solicitations to "prime" the propagation of crack, especially in these materials where the reticulum plans sliding is extremely reduced. For the UMGT order of magnitude, a single piece is composed by a low grains number, so the problem of the structure inner defects is more controllable. The piece surface-volume ratio grows, and will be, therefore, necessary to put greater attention to the superficial defects, prime points for the propagation of crack. From this first comparison, SiC and Si3N4, seem to be the more appropriate materials, thanks to the better behaviour at the high temperatures. There is Recently, adding some elements which boron, aluminium, yttrium and/or relative oxides, meaningful improvements of the mechanical properties have been verified. As previously said, the characteristics of these materials depend on the manufacturing process.

5.1 Silicon carbide (SIC)

SiC presents an optimal material specific strength, a good resistance to the thermal shocks, thanks to its relative high thermal conductivity. In Table 3 the characteristics of the SiC according to the adopted manufacturing process are listed. The data refer to temperatures between the 20°C and 1400°C and the sources include the main manufacturers companies and some scientific organs, including the NIST (National Institute for Standards and Technologies, USA). Data are, sometimes, discordant depending on the manufacturing process and operational temperature, or because the test and production techniques are not perfectly standardized. In general the worse properties obtained with the Reaction Bonding, the more economic process, and the best ones with the CVD, most expensive one. The Sintering and the Hot Pressing show analogous intermediate characteristics and not too much distant from the CVD. The numerous and reliable data in literature are reported in Table 4. A sintered SiC has been considered, and the result have been completely validated by the NIST.

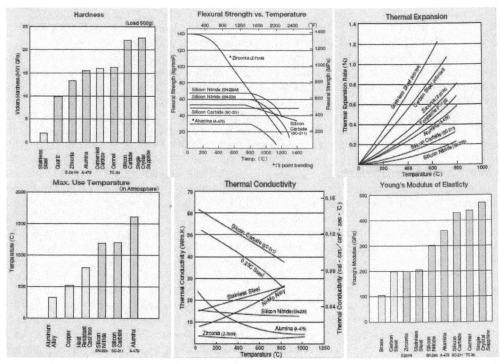

Fig. 13. Materials characteristics

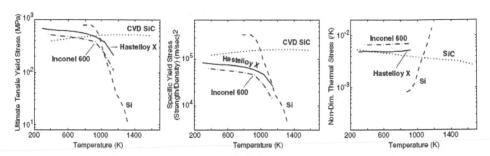

Fig. 14. Materials proprieties at varying of operational temperature [NIST]

SiC	Reaction Bonding	Sintering	CVD	Hot Pressing
Creep Rate Exponent	1.6	1.1	-	0.9
Density [g/cm³]	2.9-3.1	3.05-3.17	3.21	3.1-3.3
Elastic Modulus [GPa]	275-390	372-450	434-476	380-451
Flexural Strength [MPa]	190-400	359-511	468-575	400-500
Fracture Toughness [MPa m$^{1/2}$]	4	2.6-4	2.7-3.5	3.9
Hardness Vickers [GPa]	22-26	23-26.7	27	21
Max use Temperature °C	1350-1600	1400-1600	1400-1600	1400-1600
Tensile Strength [MPa]	77-310	234-310	220-310	200-310
Thermal Conductivity [W/m K]	110-200	31.3-116	63-300	50-120
Thermal Expansion From 0°C [10^{-6} K^{-1}]	4.3-4.6	4.2-5.9	4.0-4.6	4.3-4.6

Table 3. SiC proprieties at varying of manufacturing process

Percentage in parentheses denote estimated combined relative standard uncertaints of the propriety. For example, 3.0 (5%) is equivalent to 3.0 ± 0.15. Property values in parentheses are extrapolated					
Property [unit]	20°C	500°C	1000°C	1200°C	1400°C
Bulk modulus [GPa]	203 (3%)	197	191	188	186
Creep rate [10^{-9}s^{-1}] at 300 Mpa	0	0	0	0.004 (17%)	0.27
Density [g/cm³]	3.16 (1%)	3.14	3.11	3.10	3.09
Elastic modulus [GPa]	415 (3%)	404	392	387	383
Flexural strength [MPa]	359 (15%)	359	397	437	446
Fracture toughness [MPa m^{-2}]	3.1 (10%)	3.1	3.1	4.1	4.1
Friction coefficient[], 0.2 m/s, 5N	0.7 (21%)	0.4	0.4		
Hardness (Vickers,1 kg)[GPa]	32 (15%)	17	8.9	(6.9)	(5.3)
Lattice parameter a (polytype 6H) [Å]	3.0815 (0.01%)	3.0874	3.0950	(3.0894)	(3.1021)
Lattice parameter c (polytype 6H) [Å]	15.117 (0.02%)	15.1440	15.179	(15.194)	(15.210)
Poisson ratio []	0.16 (25%)	0.159	0.157	0.157	0.156
Shear modulus [GPa]	179 (3%)	174	169	167	166
Sound velocity, longitudinal [km/s]	11.82 (2%)	11.69	11.57	11.52	11.47
Sound velocity, shear [km/s]	7.52 (2%)	7.45	7.38	7.35	7.32
Specific heat [J/kg K]	715 (5%)	1086	1240	1282	1318
Tensile strength	250 (6%)	250	250	250	250
Thermal conductivity [W/m K]	114 (8%)	55.1	35.7	31.8	27.8
Thermal diffusivity [cm²/s]	0.50 (12%)	0.16	0.092	0.079	0.068
Thermal expansion from 0°C [10^{-6} K^{-1}]	1.1 (10%)	4.4	5.0	5.2	5.4
Wear coefficient (log10) [0.2 m/s, 5N]	- 4.0 (5%)	-3.6	-3.6	...	...
Weibull modulus []	11 (27 %)	11	11	11	11

Table 4. Sic proprieties [NIST}

5.2 Silicon nitrate (Si$_3$N$_4$)

The silicone nitrite is a material much industrially diffusing. In Table 5 the main characteristics of the Si$_3$N$_4$ are reported, according to the manufacturing process, for a temperature within 20°C and 1400°C. Data, relative to CVD operations, are not available, because for such material is only used to generate protecting films, not for solid pieces

Si$_3$N$_4$	Reaction Bonding	Sintering	CVD	Hot Pressing
Creep Rate exponent	1.7	1.1	-	1.2
Density [g/cm^3]	2.3-2.6	3.2-3.27	-	3.1-3.31
Elastic Modulus [GPa]	155-200	245-310	-	175-320
Flexural Strength [MPa]	190-338	70-703	-	146-930
Fracture Toughness [MPa m$^{1/2}$]	2-3.6	4.3-6	-	3-8
Hardness Vickers [GPa]	10	14.8-15.7	-	13.9-15.9
Max use Temperature °C	1200-1500	1000-1500	-	1200-1500
Tensile Strength [MPa]	140-170	140-576	-	140-726
Thermal Conductivity [W/m K]	10-16	26-30	-	15-42
Thermal Expansion From 0°C [10^{-6} K^{-1}]	2.9-3.3	3.1-3.5	-	2.7-4.3

Table 5. Si$_3$N$_4$ proprieties

creation. From a data analysis, the properties of the Si3N4 are strongly variable, also presenting higher values than the SiC ones. In effects the silicon nitrite has optimal characteristics to low temperature, but it degrades quickly with the temperature increase. To high temperatures, also maintaining a good behaviour, does not reaches the SiC values. In this case, it can be noticed as the Reaction Bonding does not allow to obtain good property because of the achieved lower density than the theoretical one: the porosity is too much high.

5.3 Tenacity, fracture toughness, creep, time to failure

The fragile behaviour of the ceramics focuses the study of the mechanical characteristics on the field of the mechanics of the elastic linear fracture. The design, in this case, is difficult because the microcrystalline structure of the ceramics introduces intrinsic imperfections in the mold preparation and edging processes, that render the material properties extremely variable. A greater limit of the ceramic materials is the low fracture toughness K_{IC} and, consequently, the greater probability of structure collapse due to the crack propagation. The critical crack length "l_c" represents the dimension which the phenomenon increase and becomes spontaneous and irreversible. This event is not correlated to the operational environment but linked to the produced energy during the inner material tensions displacement. This largeness represents the maximum tolerable dimension of a crack in operational conditions, and determines the structure lifetime. Beginning from the energetic criterion of Griffith [Frechette] the simplified formula can be written:

$$K_{ic} = \sigma \sqrt{\pi \cdot l_c} \tag{24}$$

Once σ and K_{IC} are known (Table 4 e 5) the SiC critical crack length can be calculated:

$$\sigma = SU_1^2 \rho_m = 133.2 \text{ MPa} \Rightarrow l_c = 172 \text{ μm} \tag{25}$$

for Si$_3$N$_4$:

$$\sigma = SU_1^2 \rho_m = 137.6 \text{ MPa} \Rightarrow l_c = 420 \text{ μm} \tag{26}$$

The silicon nitrite have a better tenacity, but of a lower value (140 MPa, Table 5) of the maximum permissible stress to traction at high temperature, value that appears insufficient to assure adequate safety margins. It can be noticed as the critical length, at this scale, is an order of magnitude lower than the blade dimensions., and two orders of magnitude higher than the device accuracy, that evidently would lose its functionality with a crack "so large". Recent studies, aimed to improve the tenacity of the ceramic materials, thanks to the addition of some additive (Al, B and C), to clearly increase the K_{IC} till values (for SiC) of 9. It can be assumed that the machine lifetime, subordinate to the maximum static stress (as previously calculated), is higher than to 10^4 hours if SiC is used (Figure 15), while is ~600 hours for the Si_3N_4 (Figure 16). In both cases the structure deformations, due to the sliding, are of little μm, therefore negligible because the design tolerances that are of about 10 μm.

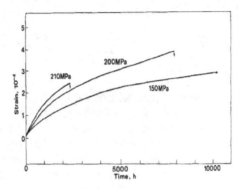

Fig. 15. Alumina SIC alloy creep curves at 1400°C [NIST]

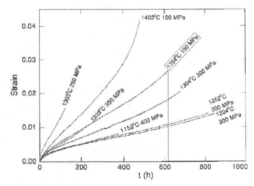

Fig. 16. Si_3N_4 creep curves [NIST]

The mechanism of complete structure failure the under a stationary cargo has had to the interaction between sliding and defects existence, to the speed of crack propagation. Both properties depending on the manufacturing process.

In static conditions, the solicitation, where, superficial carvings or micro cavity between grains, are present, generates a crack advance along the grain edges, the material weaker point. Once reached the critical length, the structure fails and this happens in a determinate time, defined Time to failure; a stress σ_{th} of minimal threshold exists, necessary to prime the process of crack propagation. Under these conditions, it seems that the piece is capable to resist for an indeterminate time. For silicon carbide it has been extrapolated (Figure 17): σ_{th} = 165 MPa to traction to 1400°C. It can be noticed that for values higher than the threshold one,

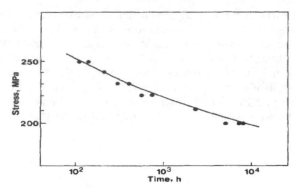

Fig. 17. SiC *Time to failure* at 1400°C [NIST]

the lifetime is maintained to 10^4 hours. For Si$_3$N$_4$ this value is lower: $\sigma_{th} \approx 150$ MPa to traction at 1300°C; Figure 18 shows how this value rapidly decrease at the temperature increasing. Particularly interesting is the fatigue behaviour, that it is one of the main mechanisms of priming and lengthening of a crack, and in for ceramic materials represents the determining factor of the lifetime. In the fragile materials is distinguished between static fatigue and cyclical: the first one is related to the supported load, with variable conditions of temperature and in oxidating atmosphere, while the second, the most meaningful one, is determined by variable solicitations conditions. The fundamental law that regulates the material fatigue lifetime is famous as law of Paris, the expresses the crack propagation velocity in "N" number of cycles:

$$\frac{da}{dN} = C\Delta K_i^{\ m} \qquad\qquad (27)$$

The constants C and "m" is closely correlated to the material, while the ΔK_i represents the variation of the stresses intensity factor (That means it is correlated to the structure applied solicitation). In the metals the exponent is small (m between 2 and 4) and an increase, as an example of a factor 2, the $\Delta\sigma$ applied to the machine reduces the component life-cycles of a order of magnitude, acceptable during design procedures.

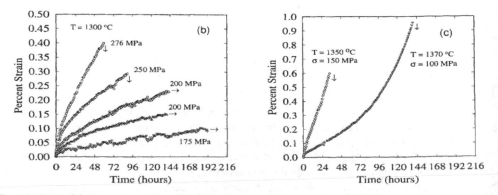

Fig. 18. Si$_3$N$_4$ lifetime at varying of solicitations and temperature [NIST]

For the ceramic materials, the exponent is much higher (m > 20) and it is enough a factor 2 to modify the he component life-cycles at least six orders of magnitude. It is possible but to take advantage, in the preliminary design phase, of the threshold value of the ΔK_i, that defines the variation of the stress intensity factor, and consequently, the value of the stress under of which the fatigue is negligible. Such value is usually around 50% of the K_{ic}; in the event of the SiC reinforced with Al, B, C the data are represented in Figure 19. Operating under this threshold value, it is possible to reduce the influence of the fatigue on the machine lifetime to a secondary parameter, even if, some factors which the corrosive atmosphere or the inner disposition of the micro cavities can limited the really usable an allowable ΔK_i. The available data for the UMGTs, actually under investigation, unfortunately, are not sufficient to determine the influence of the fatigue on the machine.

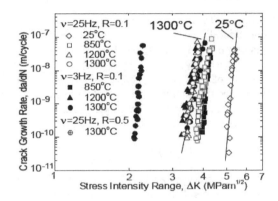

Fig. 19. Si₃N₄ Stress intensity range

5.4 Ceramic materials

The ceramic materials are interesting for structural applications as they are thermally stable and not so heavy. In fact, since between their atoms create covalent or ionic bond, such materials generally have a high chemical inertia, a high elastic modulus and a remarkable hardness, also to temperatures over 1000°C. The same chemical bonds that give them interesting characteristics are also responsible of their fragile behaviour. They do not allow to the crystalline plans to slide ones regarding the other, and do not allow the material plastic deformation. Consequently the ceramic material presents a typical fragile fracture: without some warning, with a highest crack propagation velocity. Such behaviour is described using the Griffith law:

$$\sigma = \frac{K_{IC}}{\sqrt{\pi a}} \qquad (28)$$

To render ceramic a reliable material (or better to increase its Weibull module) it is necessary or increase the value of its critical stresses intensification factor or decrease the inner defects dimensions. In the first hypothesis it needs to modify the microstructure by new insertion of "new phases" in the matrix, in the second needs to optimize composite "processing" of and to take care the superficial finishing.

5.4.1 Weibull statistic

The design of ceramic components are based on three types of approach:

- an *empiricist* approach, based on the iterative tests execution of that finish when the member satisfies the properties demanded. Sometimes, if the modelling and the previewing, this is the only solution;
- a *determinist* approach, in which it is attempted , by mathematical models, to preview the material resistance behaviour. This approach is correlated to the FEM analysis, and satisfied with the metals but sometimes it is inadequate for the ceramic ones, especially in the design loaded by critical stresses;
- a *probabilistic* approach: it is assumed that a data volume of ceramic material loaded by uniform stress breaks for effect of the defect of greater entity (approach of Weibull)

Weibull (1939) was the first one to introduce, through an statistics analysis, the concept of breach probability. In its model the member is considered as "a chain" constituted from N meshes: the failure of this chain when the destructive breakdown of a single mesh occurs, that weaker one. Such event is independent by the other possible events (failures). In statistical terms, the reliability of the chain is calculated as the product of the reliability of each mesh. The concept is perfectly suitable to the examination of a ceramic volume. To such purpose the experimental observation indicates that the Failure Probability $P_{f,u}$ is given by:

$$P_{f,u}(\sigma) = 1 - e^{-\left(\frac{\sigma}{\sigma_0}\right)^{\beta}} \tag{29}$$

in which σ_0 and β are material constants. According to this model the probability is zero for null tension and unitary for tension sufficiently elevated. The A_u reliability of the of volume element, equal to the complement to 1 of the probability, is:

$$A_u(\sigma) = e^{-\left(\frac{\sigma}{\sigma_0}\right)^{\beta}} \tag{30}$$

According to the "chain" model, the $A_{V(\sigma)}$ reliability, of a component of volume V subject to a uniform stress σ, is equal to the product of the reliabilities of the unitary elements that compose it, that is:

$$A_V(\sigma) = e^{-V\left(\frac{\sigma}{\sigma_0}\right)^{\beta}} \tag{31}$$

So failure probability is equal to:

$$P_{f,V} = 1 - e^{-V\left(\frac{\sigma}{\sigma_0}\right)^{\beta}} \tag{32}$$

To this the function corresponds to the probability density of failure, expressed by the relation (32). The defined formula is indicated as 2 parameters Wiebull distribution. The parameter β is called "Weibull modulus" while the parameter σ_0 is a scaling parameter. If the tensional state is not uniform, the reliability of the component of volume V is given by the product of the single constituent unit reliabilities of the total volume. Finally, it can be noticed that β modulus expresses a "simple probability" of the ceramic material behaviour.

$$P_r(\sigma) = \frac{\partial P_{r,V}(\sigma)}{\partial \sigma} = \frac{\beta V}{\sigma_0}\left(\frac{\sigma}{\sigma_0}\right)^{\beta-1} e^{-V\left(\frac{\sigma}{\sigma_0}\right)^{\beta}} \tag{33}$$

5.5 Monolithic materials and composites

The monolithic ceramic materials, that is without reinforced fibres are not adapt for applications to high temperature, due to the low resistance to the thermal shocks and for fragile behaviour. Since than ten years the research is being focused on ceramic materials that are constituted by ceramic reinforced matrices with ceramic fibres. The composites present a better tenacity and fracture toughness. We can notice that, also having chosen a matrix and a phase to increase its tenacity, to put together these elements and to create a composite with sufficiently low defects is not so simple. For composites with polymeric and metallic matrix, the problem is not banal: the matrix can be, in fact, brought to the liquid state, and consequently during the cross-linking and the cooling processes, the amount and the dimensions of the vacancies are rather small. Different is the case of ceramic composites. The matrix cannot be melt, because (moreover the technological difficulties linked to the manipulation of the liquid phase at temperature over 2000°C) or it is decomposed before melting or its fusion temperature are much high to react with the tenacity phase. The only method to increase the matrix density, is the sintering process, that can be defined schematically as a succession of hot transformations of the material structure that take advantage of mechanisms of gaps and species gaseous diffusion: in such a way a gas expulsion contained in green ceramic material is provoked, eliminating the excessive porosity and increasing the material compactness.

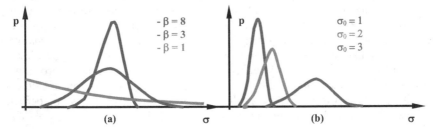

Fig. 20. β (a) and σ (β) distribution at varying of probability density

The model of the contact spheres elaborated by Frenkel and Kuczynski in 40s, explains metals and glasses densification process (the light violet sphere indicate the glass phase increase). This model is extended by Kingery to the sintering for the solid state diffusion for the ceramic materials. This process starts from the ceramic powders, to which has to be added a second phase, often constituted by particles with dimensions of some micron and lengthened shape. All it must then be heated to high temperature to prime the sintering process. It is opportune that the particles dimensions of the two phases can be comparable, to avoid that the larger particles stop or delay the process, working as "rigid inclusion" and producing a low density material with great vacancies amount, with poor mechanical properties, even if the K_{IC} has been improved by the presence of the new phase. Generally, to produce ceramic composites an "oxidic" matrix can be used, like the alumina, the zirconium, ecc. Moreover some matrix with essentially covalent bond can be used, as an example the silicone nitrite, boron nitrite or aluminum, the silicon carbide and others. Between the oxides based matrices, the alumina is most used. It has good mechanical properties, but its main propriety is the resistance to the usury and to the oxidation; less good are its creep resistance. The tenacity and the heat conductivity are lower. Other materials can be remembered, thanks of their wide use, like the yttrium, zirconium, and various carbides and nitrides. Both monolithic and composites materials, even if characterized by elevated stability, are subject to several forms of structural decay if loaded at high temperature.

5.5.1 Property of the ceramic materials and high temperature behaviour

The properties of the ceramic material at temperatures over 1600°C are analyzed here below. For the applications in study, they are:

1. oxidation resistance;
2. structural resistance to traction/compression and shear stresses;
3. creep;
4. tenacity;
5. fatigue.

5.5.1.1 Oxidation resistance

Many structural applications to high temperature demand the exposure to highly oxidant atmosphere: the thermal stability and the oxidation resistance to the oxide materials become of fundamental interest. A comparison of parabolic rates of oxidation for not oxidic ceramic materials is introduced in Figure 21. In general, an acceptable rate of oxidation is attested around 10 $\mu m^2/h$; this is equivalent to assert that the material produces 100 μm of oxidized slags in 1000 hours. As previously described, the great part of carbides, borides and nitrides, oxidize to rates that exceed of several orders of magnitude this threshold value, and consequently they can not be used in the long term structural applications to high temperature. Only the silicon nitride Si_3N_4 and carbide SiC are able to maintain the previously determinate value to the temperature of 1600 °C. Although the great part of the oxidic ceramic materials are considered stable in a full oxygen atmosphere, the phases interfaces can be oxidized in presence of an oxygen gradient; as an example, if stable ternary oxide, ABOx, are subject to an oxygen gradient and if the relative diffusion coefficients to the three elements are in the relationship $D_A > D_B > > D_0$, the crystal will rich of AO on the side of the greater gradient.

5.5.1.2 Traction/Compression and shear structural resistance

Many polycrystalline ceramic materials quickly lose their characteristics of structural resistance at high temperature. Hillig indicates that the resistance of fragile materials would have to decrease proportionally to the 3/2 of the T/T_{melt} ratio (T_{melt} is the melting temperature of the materials) and that the value would have decrease of the half approximately if the environment temperature is equal to 0.5 T_{melt}. This behaviour is the result of the diffusion mechanism control of the particles that dominate the processes of deformation to the high temperature. A schematic diagram of the resistance limits correlated to the mechanisms of fracture and creep in the ceramics is shown in Figure 24. To high temperatures many materials show ductile behaviours; this involves that their properties of resistance immediately are controlled by creep phenomena. Consequently, the advance of the micro-crack in the material is regulated by the creep. In the table 6 and Figure 23 the resistance limits and the fields of application of several ceramic materials are shown s: for many applications is necessary not to exceed tensions of the order of 10^2 MPa (typically 150 MPa) to safety reasons. No ceramic materials shown in figure seem able to support this tension to approximately 1600 °C; consequently the composites materials reinforced with fibre are the only ones that allow reaching such design specifications. The titanium boride they seem to be of the possible candidates to reinforce of oxidic matrix, having excellent resistances to the high temperatures, but highly oxidable even if they are included in the matrix.

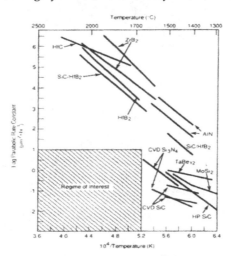

Fig. 21. Oxidation curve

Also the pure silicon carbide introduces an adapted resistance in the temperature range of interest, but the structural SiC and its fibres are typically lacking the necessary purity to obtain high levels of resistance, and consequently, present meaningful proprieties losses to 1200°C. A better quality in the realization of silicon carbide fibres is being obtained with the new formation processes to high temperature. The oxidic ceramic materials, on the other hand, do not present the necessary resistance in high temperature atmospheres. Meaningful improvements have been obtained through the increase of the hardness of the solution or the control of the precipitate (above all zirconium), but the allowable stress have been always maintained lower than 150MPa. The sapphire (mono crystalline alumina) shows the necessary degree of resistance if oriented in opportune way during its cooling, but the more amazing result have been obtained using a two-phase material apt to receive ulterior oxidic phases: the eutectic system Al_2O_3/YAG, is capable to support stress of about 250 MPa to 1650°C. Studying this material, it appears clearly, that in the ceramic the crystals orientation is most important and, therefore, the direction of solidification of the green ceramic.

5.5.1.3 Creep

Figure 24 illustrates that creep is the dominant dynamic process to the high temperature, as well as that it is assumed as design criterion in any structural application. The creep rates must be of the order of $10^{-8}/s$ or lower for long term applications; to obtain these resistant levels, a creep resistance fibre must be deposited into a less resistant matrix. It is obvious that the polycrystalline ceramic fibres are not able to supply an adequate creep resistance for such matrix. The oxidic and monocrystalline fibres analyzed by Corman seem to be more promising. As an example, it can be noticed the interesting behaviour of the mono crystalline yttrium-alumina-garnet $(Y_3Al_5O_{12})$: this fibre present an interesting creep resistance (in our application field), and, being to cubical material, its creep advancing rate is independent to the direction of its crystals; for this reason, the fibres having such structure, present a better resistance to complex stresses states.

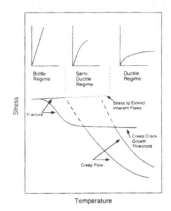

Fig. 22. Stress/temperature diagram

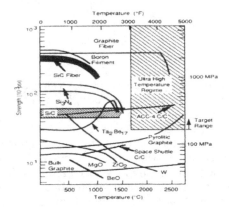

Fig. 23. Application field

Material	Temperature(°C)	Yield Stress(MPa)
Al_2O_3 - prismatic	1600	80-120
Al_2O_3 - pyramidal	1600	200-300
ZrO_2 (4.5 mol% Y_2O_3)	1400	550
ZrO_2 (9.4 mol% Y_2O_3)	1400	160
ZrO_2 (9.4 mol% Y_2O_3)	1600	40
ZrO_2 (18 mol% Y_2O_3)	1400	275
TiB_2	1600	> 800
TiB_2	1900	500
SiC	1600	350
WC	1600	100

Table 6. Resistance characteristics

It is not so easy, on the contrary, to preview the behaviour to creep of reinforced composite ceramic; a dominant parameter to take under control, in such composites, is fibre aspect ratio.

The stationary creep phenomenon on formed ceramic materials is analogous to that one shown for the metallic ones, above all, in relation to the dependency by the module of applied stress. Cannon shows how the exponent (n) in the formula (36) varies within 3 to 5 in the "dislocation climb" and within 1 to 2 when the diffusion mechanisms are active. It is assumed here that the behaviour to creep of a ceramic fibre and its matrix can be described from a common "power law" expression of the type:

$$\dot{\varepsilon} = A\sigma^n \exp\left(-\frac{Q}{RT}\right) \tag{34}$$

where the composite deformation speed $\dot{\varepsilon}_c$, and the σ_C stress, are weighed with the volume fraction of the two phases. The composed stress σ_C can be correlated to creep through the following formula:

$$\sigma_c = \frac{\sigma_{fo} V_f \dot{\varepsilon}_c^{1/n}}{\dot{\varepsilon}_{fo}^{1/n}} + \frac{\sigma_{mo} V_m \dot{\varepsilon}_c^{1/m}}{\dot{\varepsilon}_{mo}^{1/m}} \tag{35}$$

where, σ_{fo}, ε_{fo}, σ_{mo}, ε_{fo}, are obtained from material empiric relations, and "m" and "n" are the stress component for the matrix and fibres respectively. This equation allows calculating demand stresses to support a stationary creep speed in a continuous composite, being assumed that both fibres and matrix are sliding to the same speed and without distortions on the phase interfaces.

5.5.1.4 Tenacity

Before using a fragile material in a structural application the resistance limit to thermal shock, damaging from impact and fast fracture for cracking advance must be indicated. As previously shown, the creep is the process of dominant deformation when the sliding tension is lower than the necessary one to prime unstable crack advancing. The intensity of

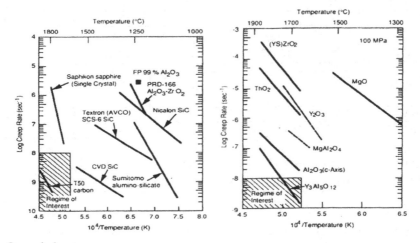

Fig. 24. Creep behaviour

the applied and localized stresses shows different increases in relation to the crack length; this produce a ductile behaviour in the material when the stress, K_m, is lower than the limit stress, K_{th}. The tenacity is the result of the applied traction on the fracture of the matrix by same fibres behind the fracture line: the friction of the fibres to the sliding of broken fibres opposes resistance in the crack matrix. The table 7 reports the studied structural ceramic composite properties.

Composite reinforcement/ Matrix	Flexure or Bend Strength (MPA)		Creep @ 100 MPa s^{-1}	R.T. Toughness MPa m$^{-1/2}$	Fatigue	Oxidation resistance
	R.T.	Elevated Temp.				
Reinforcements						
Carbon	2000-4000	≈2500 @ 2300°c	10^{-11} @ 1600°C	U	U	Catastrophic
TiB$_2$/TiB(C)	670-1560	900@ 1000°C	U	U	U	Poor
Sic (CVD)	3200	1200 @ 1400°C	3 ·10^{-9} @ 1600°C	U	U	Good
SiC (polymer)	1900	1050 @ 1400°C	5 ·10^{-9} @ 1400°C	U	U	Good
C-axis Al$_2$O$_3$	1400-3000		10^{-9} @ 1600°C	U	U	Stable
C-axis BeO	U	≈40M @ 1000°C	10^{-9} @ 1750°C	U	U	Stable
YAG	U	U	5 ·10^{-10} @ 1600°C	U	U	stable
Nonoxide/Nonoxide						
BNp/AlN-SiC		28 @ 1530°C	U	U	U	Fair to 1600°C
SiC/SiC	350-750	U	U	18	U	< 10□m²/h (1600°C)
SiC/HfB$_2$	380	28 @ 1600°C	10^{-5} @ 1600°C	U	U	12□m (2000°C)
SiC$_p$/HfB$_2$-SiC	1000	U	U	U		5% wt gain (1600°C)
SiC/MoSi$_2$	310	20 @ 1400°C	U	≈ 8	U	< 10 m²/h (1600°C)
ZrB$_2$pl/ZrC(Zr)	1800-1900	U	U	18	U	U
20vol% SiC/Si$_3$N$_4$	500	U	U	12	*	Good to 1600°C
Nonoxide/oxide						
SiC/ZrB$_2$-Y$_2$O$_3$	U	16 @ 1530°C	U	U	U	Poor
TiB$_2$/ZrO$_2$	U	U	U	U	U	Poor
SiC/Al$_2$O$_3$	600-800	U	10^{-5} @ 1525°C	5-9	U	Poor
30vol% SiC/ZrO$_2$	650	40 @ 1000°C		12	U	Poor > 1200°C
Oxide/oxide						
Al$_2$O$_3$/mullite	≈180		U	U	U	Dissolution reaction
Al$_2$O$_3$/ZrO$_2$	500-900		U	U	U	Stable
YAG/Al$_2$O$_3$	373	198 @ 1650°C	U	4	U	Stable

U = unknown
* = 0.22mm crack after 50000 cycles @ 42 MPa in compression

Table 7. Ceramic material proprieties [NIST]

5.5.1.5 Fatigue

The today's design strategies are based, first of all, on the tenacity and resistance of the material. The behaviour to cyclical fatigue, thermal and mechanical one, must be enclosed as an essential element in the design analysis. The *thermal fatigue* takes part on every material under re-heating and cooling cycles. The thermal shock is the effect and it is present every time that a material is not constantly heated or it is not held to a constant temperature. Naturally, when a material is heated in a furnace, above all if its heat conductivity is low, and is quickly heated, or vice versa, quickly cooled, it creates thermal gradient in the material. In these conditions every such material are characterized by embrittlement, thermal fatigue, micro fractures, crack, etc. that, in the long run, produces a thermo mechanical failure. These important considerations have to be mentioned, especially during the preliminary design phase. In fact it is not enough to consider the maximum temperature of resistance of a material or the mechanical resistance, but all the problem boundary conditions that indicates the better choice of the material to use. The *mechanical fatigue* is caused by "slow crack growth" (SCG) mechanisms that must be analyzed for nominal load conditions. In spite of the acquaintance of the mechanism, the device lifetime prevision is based on empiric correlations for the crack rate. Generally a "power law" is adopted, whose main advantage resides in the mathematical model simplicity:

$$v = \frac{da}{dt} = A\left(\frac{K_I}{K_{IC}}\right)^n \qquad (36)$$

where "v", "a" and "t" are the crack rate, its characteristic length and the time respectively. K_I and K_{IC} are the stress intensity factors (effective/critical). "a" and "n" are parameters depending on the material and the operational atmosphere. The lifetime (t_f) depends essentially on the applied stress (assumed constant) and on the K_{IC}:

$$t_f = D\sigma^{-n} \qquad D = f(K_{IC}; A; n) \qquad (37)$$

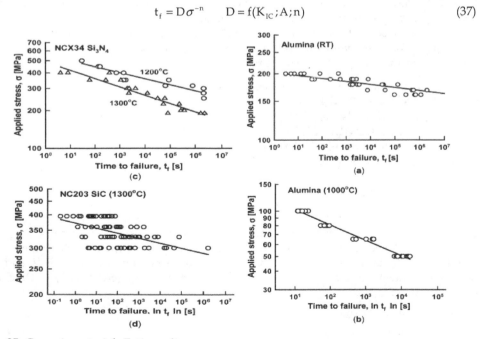

Fig. 25. Ceramic materials Fatigue diagrams

5.6 Metallic materials

5.6.1 Stainless steel alloy

Low C tenor alloy (except some martensitic steels), or better Fe-Cr or Fe-Cr-Ni alloy, with Cr percentage varying from the 12 to 30% and of Ni from the 0 to 35% (main characteristic is the corrosion resistance). Such characteristic is due to the passivation property that such steels, in oxidation conditions, have (that is capacity to create an oxide film of extremely thin dimensions whose characteristics remarkably according to change the chemical composition of the alloy, of the heat treatment, the structural composition, the superficial stress state) . Normally they are subdivided in following the three classes:

- Martensitic steel ;
- Ferrite steel;
- Austenitic steel.

The martensitic steels have the critical points and are submitted to heat treatment; they have higher mechanical characteristics in comparison to the ferritic and austenitic ones, but lower corrosion resistance. The ferritic steels, instead, do not possess the critical points, can only be submitted to re crystallization annealing treatments; they have good resistance to oxidation at high temperature, as well as more elevated how much greater are the chromium content. the austenitic steel are, between the stainless steel, those mainly produced. They have elevated corrosion resistance in numerous atmospheres; they are submitted to heat volatilization treatment (hardening of austenitic steel) to melt, in the austenitic matrix, the Cr carbides (their presence decrease the corrosion material resistance). For this same reason the C tenor has to be held to low values. Besides these three classes are the stainless steel hardening by precipitation, with high mechanical characteristics, whose corrosion resistance, also being lower than the austenitic types, is higher than martensitic ones. Hardening process derives by the precipitation insoluble phases dispersed during the aging process. Between this type of steels we remember:

1. quality steels for mechanical constructions (UNI 3158);
2. high tensile steels (UNI 4010);
3. usury resistant steels (UNI 3160);
4. heat resistant steels (UNI 3159);
5. corrosion resistant steels (UNI 3161);
6. creep resistant steels (UNI 3608).

5.6.2 Refractory alloy

To this group belong the high temperature resistant metallic materials. They are used in gas turbines manufacturing, reaction engines, inner parts of furnaces. They must present a good corrosion resistance at high temperature (the oxidation phenomena at high temperature) and be capable to support attacks from combustion ashes (in particular S, V and its derivates, contained in the ashes), and good creep strength (creep-fluage). They are distinguished in:

- Fe base alloy;
- Ni base alloy;
- Co base alloy.

The refractory steels are used for those products whose main characteristic is the resistance to hot gases and residual of combustion, at higher temperature than 550 °C. In European code for these materials is EURONORM 95.

5.6.3 NICHEL alloy (Ni)

Melting point is approximately 1455 °C. The Ni, in steels alloy, reduces the grain dimensions, increases the depth of hardening and renders them less sensitive to the low temperatures fragile failure. The Ni can be bond with numerous metals (particularly with Fe, Cr, Cu). The Ni- Fe alloys can be:

- With low expansion coefficient or controlled expansion as the Invar; these alloys find applications in the fabrication of physical measurement device, bimetals and equipment for lowest temperatures;
- alloy with constant elastic modulus used for clocks springs, frequency regulators;
- alloy with special magnetic characteristics (non-magnetic, low and high constant permeability);
- Magnetic resistance alloy, for transducers;
- Cr or Co alloy, resistant to the heat;
- Cu alloy, resistant to sea water and to many chemical compounds (Monel-Metal).

5.6.4 COBALT alloy (Co)

It is used in metallurgy with Fe, Ni, Mo, W and other metals for the production of refractory alloys, steels for tools and alloys for permanent magnets. It is a constituent of maraging steels.

5.6.5 TITANIUM and TITANIUM alloy (Ti)

Ti is recently used in the industrial manufacturing but its utilization still is limited, due to the high fabrication cost. The main properties of this metal and its alloys are the oxidation resistance also to 400-450 °C and higher mechanical characteristics both cold and hot environment. Its low density (4,54), has as consequence an high σ_m/p.sp ratio. The melting point is 1672 °C. These characteristics of pure Ti and its alloys are a lot influenced by the inner contained gas and, in particular, by the H. The value does not have to exceed the 0,0125% to avoid a dangerous embrittlement. The Ti alloys find application in the fabrication of supersonic vehicles, the aerospace constructions and in particular in the missiles.

6. Manufacturing technologies and process

The UMGT design is strongly limited by the available technologies. The micro manufacturing introduces a several problems in the development of every single part of the device. All studies have bring to adopt as manufacturing techniques the DRIE (Deep Reactive Ion Etching) and the Wafer Bonding, diffuse and by now widely mature, add to Micro Reaction Sintering, a new process based on the HIP (Hot Isostatic Pressing), already experimented for the fabrication of micro turbines and capable to realize machines with an accuracy of the order of the µm, in small series and with contained production costs.

6.1 DRIE (Deep Reactive Ion Etching)

This technique essentially consists in etching wafer of base material, usually silicon microcrystalline, through a plasma beam. Considering that a wafer has a diameter between 100 and 300 millimetre, it is possible, at the same time to produces numerous devices, in parallel, as shown in Figure 26.

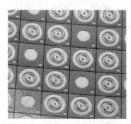

Fig. 26. Series of nano devices on a single wafer [Epstein 1997]

Initially the wafer is covered by an oxide and by a photo resistant film, on which a so-called "mask" is transferred through optical methods, a model in black and white with the geometry to recording. The transfer is realized bombing with ultraviolet rays a glass plate to contact directed with the wafer, on which the mask is applied. Geometry is so "developed" on the photo resistant material as if it were a normal photographic film. Finally and the piece is "cooked". After the "development", some oxide exposed parts remain, that are removed using solvents, leaving some area covered by the mask and some zones uncovered by silicon material. The piece is, then, etched, but the speed of the oxide etching is 50-100 times lower than of the silicon one. At the end of the procedure, the extruded initial

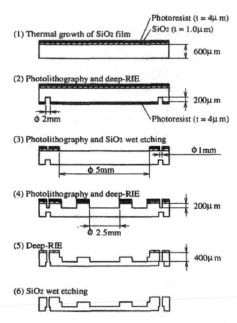

Fig. 27. DRIE manufacturing process phases

geometry is obtained. Several materials and techniques exist to produce the photo resistant mask, and SiO_2 is the most diffuse used oxide. The etching plasma beam can be composed by ion mixtures of varied nature. The process is described in Figure 27. Numerous manufacturing companies have developed particular and sophisticated machine capable to deeply etching with extreme precision, even if the adopted process is usually the Bosch patented one. Moreover several machines and device commercialized from the Alcatel, are capable to achieve an etching speed of 20 µm/min with aspect ratio of 60:1, but with the possibility to arrive till 100:1 with an ulterior cost increase. Practically, without excessive expenses, the stamps for the turbine in less than an hour and for the compressor in about 30 min could be recorded, without the necessity of successive manufacturing, since this technique has an accuracy of the order of the µm. Add the possibility to obtain tapered shovels and etchings of about 0,25 µm. The disadvantage of this technique is that simply extruded geometries can be only obtained, but already is being worked to create masks in greys tonality to producing more complex forms and shape, that could allow to have a height of the channels variable, with the consequent possibility to adopt thin profiles for the turbo machines.

6.2 Wafer bonding

This technique allows to create complex structures joining, using different other materials which the glass, wafers of already micro etched silicon. The fusion between the several parts is superficially obtained by cooking the parts, in presence of a strong cross-sectional electrostatic field that crosses the pieces interface. This system creates an alloy; the high temperatures strengthen the bonds (anodic fusion). In alternative, it is possible to use a melting process (direct fusion) in presence of modest pressure with heat transfer or it can use adhesive materials that joins the parts through superficial chemical reactions. The first two techniques create strong covalent type bonds, but they have the disadvantage to demand extremely smooth and clean surfaces for being effective, as the first step for a good resolution of the fusion is to bond the surfaces by the Van der Waals intermolecular forces. The product quality is correlated to the process temperature, between the 800°C and the 1200°C, levels of equal resistance than the crystalline structure of the same material are achieved, but already between 200°C and the 400°C sufficiently strong bonds are realized by the chemical activation of surface reactions. To improve the fusion a solution obtained from the combination of the several procedures can be used. The alignment of the layers, thanks to laser techniques, reaches precisions of the order of the µm, even if, currently, it is not possible to align more than 5-6 wafer maintaining such accuracy.

6.3 Micro Reaction Sintering (MRS)

This manufacturing techniques is developed for the production of SiC turbomachines rotors of 5÷10 millimetre of diameter. From graphite powders, α-SiC, silicon phenol resin is inserted in a stamp, derived from silicon wafer. The parts of the stamp are melted and all component is submitted to high temperature HIP. The technique has been calibrated in several conditions of pressure, temperature and composition of base powders and it has been verified that the microcrystalline structure of the final piece has an higher quality if the furnace temperatures is about 1500-1700°C, the pressure varies from 100 to 50 MPa and using α-SiC powders instead of silicon one. A microscopic analysis of the structure has demonstrated that melted stamp reacting with the graphite substrate forms a covering of β-SiC around a nucleus of α-SiC deriving from initial powders. But is not still available mechanical tests on the material. The adoption of this technique wants substantially to reduce the excessive long times of bonds production of the CVD or the DRIE on SiC wafer, caused by the speed of material deposition and recording, and to the lack of accuracy of the EDM (Electro Discharge Machining).

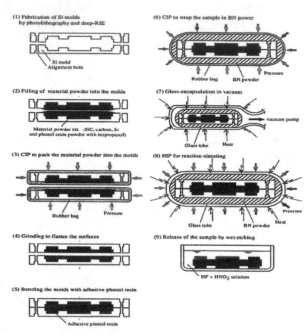

Fig. 28. Micro Reaction Sintering phases

7. Economic overview analysis and potentialities

All possible application and the alternatives of the proposed device must be investigated. It appears obvious that a UMGT will be utilized in the two main area, already covered by the large scale machines:

- the power generation
- the vehicle propulsion.

The primary application filed will be the first one; the higher energy density respect to a chemical battery with an overall efficiency of the 5-10% is sufficient to render a UMGT more attractive on the market. Substantially fuel tank is the bigger component, for which reason, in the developments of such machine, the researchers have to obtain low consumption than to high power output, thanks to the possibility to satisfy the power requirements through "modular clusters". Besides the possibility to feed every type of portable device, a such generator could go satisfy the power requirements of aircraft onboard systems, in terms of redundancy and higher operational temperatures, in comparison to the conventional battery packages. The competition with the large scale power plant is not possible, because is not imaginable of achieve efficiency of the order of 50-60%. The situation would be various in the moment in which will be possible to obtain, for example with regeneration, efficiencies about 20-30%. So the lower performances could be compensated with the advantage of the redundancy, the lowest noisiness and the UMGTs compactness. Such levels are still the much far being reached. On the other hand, in the emergency applications, where the efficiency is not a limiting requirement regarding the compactness, succeeding to contain the costs, a UMGT would not have rivals. Three requirement that render this device a lot attractive are therefore, compactness, redundancy and lower noisiness thanks to the speed frequencies of hundred of kHz (beyond the audibility threshold) and to the length small scale (the exit flow is quickly stirred with the surrounding atmosphere). In propulsion system this device is considered as one of the more attractive relatively to little reconnaissance planes (100-1000 g of takeoff weight), thanks to the high thrust/weight ratio, which scale with the inverse of the length. The ICE motors currently under construction for these applications, are ten times greater, much noisy and with greater fuel consumptions. The possibility to use UMGT cluster, for the propulsion of great aircrafts is remote, as it seems improbable to succeed to obtain lower consumption to performance parity, without to count eventual difficulties of assembling process, due to the elevated number of necessary devices (at least 10^5). The determining factor in the diffusion of a UMGT system resides , however, in the production cost. In fact, the production of this devices presents high manufacturing cost, tied to the complex technologies to adopt, amortizable with the production in large series scale. According to an esteem of the MIT (a CMOS silicon wafer of 200 of millimetre of diameter can cost \$ 600-1000), if produced in large scale, a wafer of MEMS generators could produce several kW to the cost of several thousands of dollars for the fabrication. The result is a specific cost of 0,5-5 \$/W. Other considerations have to be done for civil application and for higher power rate (within the UMGT power rate). Here below it is discussed a practical case study.

8. User definition

8.1 UMGT advantages

The main advantage of these systems compared to the existing large scale devices resides in the highest power density. Although such systems have an overall efficiency still lower, the produced power for weight unit is the higher than which will be produced by the battery systems. In fact, at the moment, the battery package power density is about 100-150 Wh/kg. Even if an increase of the power density can be presupposed, the power developed by a UMGT is, undoubtedly, higher for weight and volume unit. Also with a lower efficiency the UMGTs have a power density that is, at least, equivalent to that theoretically produced by a lithium battery package It will be noticed, then, as in the field of the same micro turbines, there is a difference, in terms of performances, according to the used fuel: for instance, the combustion of the hydrogen, in fact, concurs to remarkably raise the value of the power density regarding the obtainable result with fuel like propane or kerosene. An ulterior advantage is the operating flexibility that offers such systems: while the battery package, once exhausted theirs energetic potential, need of a period, more or less long, to recharge them, the micro turbines allow the fuel change in a very short period.

8.2 UMGT disadvantages

The main disadvantage, respect fuel cells or the battery package, is the high operational temperature. In fact, to achieve an higher output power is necessary to realize high TIT. Because of the small scale of the machine, this design constraint leads all the system to raise own average temperature: this produces high infrared radiation emissions, that, especially in military applications, cannot be tolerated. A similar problem is found in the fuel cells in which, a lower value of the operational average temperature is reached. That infrared emissions is not the only problem for the UMGT. The high outlet temperatures of the exhaust gas, can produce problems during the operational life of the machine. To obviate to such disadvantage a regenerator, even built-in would have to be installed inside the device, to cooling the UMGT components using Peltier effect.

8.3 UMGT potential application

Multiple applications can be easy found for UMGT systems. In general terms, these ones can be divided in two groups: applications as portable power generation systems or as range extender/prime mover in different vehicles (wheel vehicles, aircraft, ecc).

8.3.1 Portable power generation

Regarding this use it can be distinguished military uses and civil applications, even if, these last one, have very limited use possibility.

8.3.1.1 Military applications

The power demand of a soldier during a mission is remarkably increased during the last few years due to the adoption of new technologies. At the same time we have the necessity to limit the most possible equipment weight for the unit. In this case, use an UMGT could help in the reduction of the transported load and, at the same time, satisfy the power demand. Some of these applications are shown in figure 29.

Fig. 29. The "future" soldier

We can distinguish three various regimes for the military applications:

$$\text{medium power} \begin{cases} \text{20W, with peaks of 50W,} \\ \text{100W, with peaks of 200W} \\ \text{1000W, with peaks of 5000W} \end{cases}$$

The first group of device generates the required power for computer, radius and sensors. The second one finds application in the laser beam and the conditioning of the soldiers (uniform ventilation in external extreme climatic conditions); with this powers range the loading of batteries is possible too. At last, 1-5 kW could be used for feeding of exoskeleton : these are robotic devices that allow to reduce the payload that weighs on the soldier.

8.3.1.2 Civil applications

The UMGT civil applications, as portable power generators, undoubtedly are limited, due to the high system temperature. In fact, in this case the exhaust gas temperature is very high, and the customer safety is a primary objective in the design of the device and its utilization. However, some civil applications can be considered, as battery recharger for mobiles or for wireless tools. Also as micro cogeneration unit at small scale.

8.3.2 MAV

The main application in this field resides in the so-called the MAV (micro aerial vehicles). Their main use is, once again, of military character. The power demanded by the group is lower because the considered vehicle mass is lower than 50 g with cruise speed between the 10 and 20 m/s: in the cruise regime the shaft demanded power is about 2.5 kW, while this doubles during the phase of takeoff and manoeuvres. Due to the high heat exchange during the flight and the low required power, there are not problems regarding the infrared emissions. Such systems have already been used in the Balkans, Afghanistan and Iraq during missions of strategic character.

8.3.3 Drones and UAV

An unmanned aerial vehicle (UAV) is a machine which functions either by the remote control of a navigator or pilot or autonomously, that is, as a self-directing entity. Their largest use is within military applications. To distinguish UAVs from missiles, a UAV is defined as a powered, aerial vehicle that does not carry a human operator, uses aerodynamic forces to provide vehicle lift, can fly autonomously or be piloted remotely, can be expendable or recoverable, and can carry a lethal or non lethal payload". UAVs typically divided in six functional categories (although multi-role airframe platforms are becoming more prevalent):

- Target and decoy – providing ground and aerial gunnery a target that simulates an enemy aircraft or missile
- Reconnaissance – providing battlefield intelligence
- Combat – providing attack capability for high-risk missions (see Unmanned combat air vehicle)
- Logistics – UAVs specifically designed for cargo and logistics operation
- Research and development – used to further develop UAV technologies to be integrated into field deployed UAV aircraft
- Civil and Commercial UAVs – UAVs specifically designed for civil and commercial applications

They can also be categorized in terms of range/altitude and the following has been advanced as relevant at such industry events as Unmanned Systems forum:

- Handheld 600 m altitude, about 2 km range
- Close 1,500 m altitude, up to 10 km range
- NATO type 3,000 m altitude, up to 50 km range

- Tactical 5,500 m altitude, about 160 km range
- MALE (medium altitude, long endurance) up to 9,000 m and range over 200 km
- HALE (high altitude, long endurance) over 9,100 m and indefinite range
- HYPERSONIC high-speed, supersonic (Mach 1–5) or hypersonic (Mach 5+) 15,200 m or suborbital altitude, range over 200 km
- ORBITAL low earth orbit (Mach 25+)
- CIS Lunar Earth-Moon transfer
- CACGS Computer Assisted Carrier Guidance System for UAVs

8.3.4 Range extender in hybrid vehicle

In the last decade, governmental incentives and the ever stricter emissions regulations have prompted some of the largest world automakers to dedicate resources to the study, design, development and production of hybrid vehicles, which offer undisputed advantages in terms of emissions and fuel consumption with respect to traditional, reciprocating internal combustion engines. In fact, hybrid engines are substantially smaller than conventional ICE, because they are designed to cover the vehicle's "average" power demand, which ensures proper traction for about 99% of the actual driving time, and is exceeded only for prolonged mountain drives and instantaneous accelerations. When excess power is needed above this average, the hybrid vehicle relies on the energy stored in its battery pack. Hybrid cars are often equipped with braking energy recovery systems that collect the kinetic energy lost in braking, which would be dissipated into heat otherwise, and use it to recharge the battery. Smaller sizes and an (almost) constant operational curve lead to lower emissions. Moreover, a hybrid vehicle can shut down completely its gasoline engine and run off its electric motor and battery only, at least for a limited operational range: this "mixed operation" increases the net mileage and releases a substantially lower amount of pollutants over the vehicle lifetime. The most popular hybrid vehicles (HV) are mostly passenger hybrid cars equipped with a traditional ICE and an electric motor coupled in parallel. The thermal engine is sized, with some exceptions, for the average power, and the surplus power needed during rapid acceleration phases is supplied by the electric motor.

9. Distance between state of art and goals

From the analysis of the several prototypes and from the documentation available in literature and on the web it can be noticed as all the groups of search are to the state of prototype. Regarding the prototype of the MIT they have been found problems of rotors failure in operating conditions. The group design by the Belgian researchers seem to be at the more advanced state, but they are not available data of any application. The Japanese model has been introduced to meeting a NATO and lacks recent news. For the Italian model the problem, currently resides in leading the system to a rotational speed, by means of generator/ electric motor, sufficient to prime the combustion. Problems have been found in the bearings behaviour and their limited lifetime

10. Conclusions

Here follow the several problems that affect this technology have been listed, as eventually exceeding these design, manufacturing and realization problems, for a device mass production.

a.   The thermodynamic cycle

It is a "standard" Brayton cycle, usually adopted in large scale turbogas. With the exception of the industrial cycles, for constructive reasons, the compression ratio is extremely low in this UMGT ($p_2/p_1 \approx 2$), and therefore the total efficiency of the machine is corresponding low (7-8%). Increase of the compression ratio, recovery of the heat of the gas are possible modifications under investigation, to increase such efficiency: but such upgrading appear difficult to implement on an UMGT device.

b.   The fuel choice

The fuel used, suggested, chosen for all prototype is a liquid hydrocarbon (pentane-butane or similar). The kerosene use or other jet-fuel appears perfectly compatible. For the operational prototype, the possibility is being studied to use methane or hydrogen, but this last one generates problems of tank design (high pressure, low power density).

c.   Design problems

The analysis of existing prototypes indicates as the main design problems that can be previewed to this point are the following ones:

1.   combustion chamber: low   residence times (flammability threshold); mixing device; resistance to the high temperatures;
2.   bearings: reliability, duration (to the highest rotational speed);
3.   rotors: radial and bending/torsion instability, resistance to the high temperatures;
4.   electric generators: practical feasibility and reliability, efficiency problems;
5.   controls devices: stability, reliability;
6.   thermo-fluid dynamic analysis: reliability of the design procedures, based on experiences achieved exclusively on large scale devices. The first simulations have evidenced the necessity to adopt scale factors in the turbomachines design.

d.   Technological and manufacturing problems

They are generally correlated to the availability of opportune high resistance materials (high thermo mechanical stresses) as well as the combustion chamber, turbine, regenerator, ecc. And to the productive technologies. While such problems will not have great influence on the realization of the prototypes, an engineering study to pass "to the productive" part (post-prototypal phase) will be necessary. Finally, the economic impact of these devices will be dependent on the performance levels and the manufacturing costs, both of which have yet to be proven. It is certainly possible, however, that UMGTs may, one day, be competitive with conventional machines in a cost per installed kilowatt. Even at much higher costs, they will be very useful as compact power sources for portable electronics, equipment, and small vehicles.

11. References

[1] A.H. Epstein, S.A. Jacobson, J.M. Protz, L.G. Frechette, *"Shirtbutton-sized gas turbines: the engineering challenges of micro high speed rotative machinery"*; Proc. 8th Int. Symposium. on Transport Phenomena and Dynamics of Rotating Machinery (ISROMAC-8), Honolulu HI, March 2000.

[2] A.H. Epstein, *"Millimetre-scale, MEMS gas turbine engines"*; Proc. ASME Turbo Expo 2003; June 2003, Atlanta USA.

[3] J. Shepherd, *"Principles of Turbomachinery"*, J. Macmillan Pub. Co, NY 1956.

[4] V. Zucrow, Hoffman *"Gas dynamics I, Vol.1"*.

[5] R. Capata, E. Sciubba *"A study on thermodynamic feasibility and possible utilization of nano gas turbines"*, ICAT 2005, Istanbul, May 2005.

[6] R. Capata, E. Sciubba: IMECE2007-42118: *"Design and Performance Prediction of a Ultra-Micro Gas Turbine for Portable Power Generation"* Proceedings of IMECE2007, November 12-17, Seattle, Washington, 2007.

[7] R. Capata, E. Sciubba: IMECE2009-12281: *"The α-prototype of an ultra-micro-gas turbine at the University of Roma 1. Final assembly and tests"*. Proceedings of IMECE2009, November 11-16, Orlando , Florida, 2009

[8] C.M., Spadaccini Mehra A., Lee J., Lukachko S., Zhang X., Waitz I.A. *"High Power Density Silicon Combustion System for Micro Gas Turbine Engines"*, Paper GT-2002-30082, ASME International Gas Turbine Institute TURBO EXPO '02, Amsterdam, The Netherlands.

[9] S. Kang, J. P. Johnston, T. Arima, M. Matsunaga, H. Tsuru, F. B. Prinz, *"Micro-scale radial-flow compressor impeller made of silicon nitride-manufacturing and performance"*, Proc. ASME Turbo Expo June 2003, GT2003-38933.

[10] Balje, O., *"Turbomachines"*, J. Wiley & Sons, 1981.

[11] Moon, H. S., Choi, D., Spearing, S. M. *"Development of Si-SiC Hybrid Structures for Elevated Temperature Micro-Turbomachinery"*, J. Micro-electromechanical systems, vol. 13, no. 4, Aug 2004.

[12] Munro, R. G., *"Material Properties of a Sintered a-SiC"*, J. Physical and Chemical Reference Data, Vol. 26, pp. 1195-1203 (1997), American Chemical Society.

[13] Peirs, J., Reynaerts, D., Verplaetsen, F., Norman, F., Lefever, S. *"Development of a Micro Gas Turbine for Electric Power Generation"*, BE Paper Eurosensors 2003.

[14] Sciubba E., *"Turbomachinery: notes"*, 2001Euroma, Italy.

[15] Sugimoto, S., Tanaka, S., Li, J. F., Watanabe, R., Esashi, M. *"Silicon Carbide Micro-reaction-sintering Using"*.

[16] www.alcatel.com, *"'I-Speeder' the future DRIE etching tool for MEMS"*, April 2000, Bosch, Alcatel, PerkinElmer.

[17] Frechette L. G., Jacobson S. A., Breuer S. K., Ehrich F. F., Ghodssi R., Khanna R., Wong C. W., Zhang X, Schimdt M. A., Epstein A. H. *"Demonstration of a micro-fabricated high-speed turbine supported on gas bearings"*, BE Paper Eurosensors 2003.

[18] Jacobson S. A., Epstein A. H., *"An informal survey of power MEMS"*, 2003, Inter. Symp. On Micro-Mech. Engin., Dec. 2003.

[19] ATIP Scoop, Japan Office, *"Micro Gas Turbine Development"*, Feb 2005.

[20] Esashi M., *"Microsystems by bulk micromachining"* .

[21] www.mems-exchange.org/catalog/P3271/

[22] Abbott I. H., *"Theory of wing sections"* 1959, Dover Publ. Inc., New York.

[23] Song B., *"Experimental and Numerical Investigation of optimized high-turning Supercritical Compressor Blades"*, PhD dissertation in Mechanical Engineering at Virginia Polytechnic Institute 2002.

[24] Butler R. J., Byerley A. R., VanTreuren K., Baughn J. W., *"The effect of turbulence intensity and length scale on low-pressure turbine blade aerodynamics"*, Int. J. Heat and Fl. Flow 22 (2000) pp. 123-133.

[25] Choi J., Teng S., Han J., Ladeinde F., "Effect of free-stream turbulence on turbine blade heat transfer and pressure coefficients in low Reynolds number flows", Int.. J. Heat Mass Transfer 47 (2004) pp. 3441-3452.

[26] Park T. S., Sung H. J., "A nonlinear low-Reynolds number k-epsilon model for turbulent separated and reattaching flow", Int. J. Heat Mass Transfer Vol.38, No 14, pp 2657-2666 , 2003.

[27] Gaydamaka I. V., Efimov A. V., Ivanov M. Ja., Ivanov O. I., Nigmatullin R. Z., Ogarko N. I., "Some Aerodynamic Performances of Small Size Compressor and Turbine Stages", Proc. Int. Gas Turbine Congress 2003 Tokyo.

[28] Waits C. M., Modale A., Ghodssi R., "Investigation of gray-scale technology for large area 3D silicon MEMS structures", J. Micromechanics and Micro-engineering, 13 (2003) pp. 170-177.

[29] Lin C. C., Ghodssi R., Ayon A. A., Chen D. Z., Jacobson S., Breuer K., Epstein A. H., Schimdt M. A., "Fabrication and Characterization of a Micro Turbine/Bearing Rig", 1999.

[30] R.A. Van Den Braembussche, A.A.Islek, Z. Alsalihi: "Aerothermal Optimization Of Microgasturbine Compressor Including Heat Transfer", Proc.Int. Gas Turbine Congress 2003, Tokio.

[31] Ishihama, Y. Sakai, K. Matsuzuki, T. Hikone"Structural Analysis Of Rotating Parts Of An Ultra Micro Gas Turbine" Proc. Int. Gas Turbine Congress 2003 Tokio, November 2-7, 2003.

[32] I.V. Gaydamaka, A.V. Efimov, M. Ja. Ivanov, O.I. Ivanov, R.Z. Nigmatullin, N.I. Ogarko Turbine Department: "Some Aerodynamics Performances Of Small Size Compressor And Turbine Stages" Proc. Int. Gas Turbine Congress 2003 Tokio November 2-7, 2003.

[33] J.P. Johnson, S. Kang, M. Matsunaga, H. Tsuru, F.B. Prinz,: "Performance Of Micro-Scale Radial Flow Compressor Impeller Made Of Silicon Nitride". Proc. Int. Gas Turbine Congress 2003 Tokio, November 2-7, 2003.

[34] K. Matsuura, C. Kato,H. Yoshiki, E. Matsuo, H. Ikeda, K. Nishimura, R. Sapkota: "Prototyping Of Small-Sized Two Dimensional Radial Turbines" Proc. Int. Gas Turbine Congress 2003 Tokio, November 2-7, 2003.

[35] T. W. Simon, N. Jiang: "Micro -Or Small-Gas Turbines" Proc. Int. Gas Turbine Congress 2003 Tokio, November 2-7, 2003.

[36] Iwai: "Thermodynamic Table For Performance Calculations In Gas Turbine Engine" Proc. Inte. Gas Turbine Congress 2003 Tokio, November 2-7, 2003.

[37] M.A. Schmidt: "Technologies For Microturbomachinery"

[38] L.G. Frechette; S.A. Jacobson, K.S. Breuer, F.F. Ehrich, R. Ghodssi, R. Khanna, C.W. Wong, X. Zhang, M.A. Schmidt, A.H. Epstein, "Demonstration Of A Microfabricated High-Speed Turbine Supported On Gas Bearings" , Proc. Int. Gas Turbine Congress 2003 Tokio, November 2-7.

[39] A.H. Epstein, S.D. Senturia, O. Al-Midani, G. Anathasuresh, A. Ayon, K. Breuer K-S Chen, F.F. Ehrich, E. Esteve, L. Frechette, G. Gauba, R. Ghodssi, C. Groshenry, S.A. Jacobson, J.L Kerrebrok, J.H. Lang, C-C Lin, A. London, J. Lopata, A. Mehra, J.O. Mur Miranda, S. Nagle, D.J. Orr, E. Piekos, M.A. Schmidt, G Shirley, S.M. Spearing, C.S. Tan, Y-S Tzeng,I.A. Wait: "Micro-Heat Engines, Gas Turbines, And Rocket Engines -The Mit Microengine Project" 28th AIAA Fluid Dynamics Conference, 4th AIAA Shear Flow Control Conference, June29-July 2,1997, Snowmass village CO.

[40] Epstein,: *"Millimetre scale, MEMS Gas Turbine Engine"* Proceedings of ASME Turbo Expo 2003 Power for Land, Sea, and Air, June 16-19, 2003, Atlanta, Georgia, USA GT-2003-38866

[41] Liu, T.F. Enrich, L. Ho, H Li, S. Jacobson, Z.S. Spakovszky: *"High Speed Micro-Scale Gas Bearings For Power MEMS"*, MTL Annual Report 2005.

[42] Epstein, S.A. Jacobson, J.M. Protz, C. Livermore, J. Lang, M. Schmidt: *"Shirtbutton-Sized, Micromachined, Gas Turbine Generators* 39th Power Sources Conference, Cherry Hill, NJ,June 2000.

[43] L-A Liew, W. Zang, L. An, S. Shah, R. Luo, Y. Liu, T. Cross, M. L. Dunn , V. Bright, J.W. Daily, R. Raj: *"Ceramic MEMS- new Materials, Innovative Processing and Future Applications"* American Ceramic Society Bulletin, Vol 80, No.5.

[44] S. Martinez Romero, *"Multipoint Optimisation Of A 2d Compressor With Splitters For A Micro-Gas Turbine Application"* 2003.

[45] A.H. Epstein, S.D. Senturia, G. Anathasuresh, A. Ayon, K. Breuer K-S Chen, F.F. Ehrich, G. Gauba, R. Ghodssi, C. Groshenry, S.A. Jacobson, J.H. Lang, C-C Lin, A. Mehra, J.O. Mur Miranda, S. Nagle, D.J. Orr, E. Piekos, M.A. Schmidt, G Shirley, S.M. Spearing, C.S. Tan, Y-S Tzeng, I.A. Waitz: *"Power MEMS And Microengines* IEEE Transducers '97 Conference, Chicago, IL, June 1997.

[46] R.A. Van den Braembussche, web paper: *"Thermo-Fluid-Dynamic Design Of Ultra Micro Gas-Turbine Components"*.

[47] Web: *MEMS GUIDE* (http://www.memsnet.org/mems/processes/).

Energy, Exergy and Thermoeconomics Analysis of Water Chiller Cooler for Gas Turbines Intake Air Cooling

Rahim K. Jassim[1], Majed M. Alhazmy[2] and Galal M. Zaki[2]
*[1]Department of Mechanical Engineering Technology, Yanbu Industrial College,
Yanbu Industrial City,
[2]Department of Thermal Engineering and Desalination Technology,
King Abdulaziz University, Jeddah
Saudi Arabia*

1. Introduction

During hot summer months, the demand for electricity increases and utilities may experience difficulty meeting the peak loads, unless they have sufficient reserves. In all Gulf States, where the weather is fairly hot year around, air conditioning (A/C) is a driving factor for electricity demand and operation schedules. The utilities employ gas turbine (GT) power plants to meet the A/C peak load. Unfortunately, the power output and thermal efficiency of GT plants decrease in the summer because of the increase in the compressor power. The lighter hot air at the GT intake decreases the mass flow rate and in turn the net output power. For an ideal GT open cycle, the decrease in the net output power is ~ 0.4 % for every 1 K increase in the ambient air temperature. To overcome this problem, air intake cooling methods, such as evaporative (direct method) and/or refrigeration (indirect method) has been widely considered [Cortes and Williams 2003].

In the direct method of evaporative cooling, the air intake cools off by contacts with a cooling fluid, such as atomized water sprays, fog or a combination of both, [Wang 2009]. Evaporative cooling has been extensively studied and successfully implemented for cooling the air intake in GT power plants in dry hot regions [Ameri *et al.* 2004, 2007, Johnson 2005, Alhazmy 2004, 2006]. This cooling method is not only simple and inexpensive, but the water spray also reduces the NOx content in the exhaust gases. Recently, Sanaye and Tahani (2010) investigated the effect of using a fog cooling system, with 1 and 2% over-spray, on the performance of a combined GT; they reported an improvement in the overall cycle heat rate for several GT models. Although evaporative cooling systems have low capital and operation cost, reliable and require moderate maintenance, they have low operation efficiency, consume large quantities of water and the impact of the non evaporated water droplets in the air stream could damage the compressor blades [Tillman *et al* 2005]. The water droplets carryover and the resulting damage to the compressor blades, limit the use of evaporative cooling to areas of dry atmosphere. In these areas, the air could not be cooled below the wet bulb temperature (WBT). Chaker *et al* (2002, 2003), Homji-meher *et al* (2002) and Gajjar *et al* (2003) have presented results of extensive theoretical and experimental studies covering aspects of fogging flow thermodynamics, droplets evaporation, atomizing nozzles design and selection of spray systems as well as experimental data on testing systems for gas turbines up to 655 MW in a combined cycle plant.

In the indirect mechanical refrigeration cooling approach the constraint of humidity is eliminated and the air temperature can be reduced well below the ambient WBT. The mechanical refrigeration cooling has gained popularity over the evaporative method and in KSA, for example, 32 GT units have been outfitted with mechanical air chilling systems. There are two approaches for mechanical air cooling; either using vapor compression [Alhazmy (2006) and Elliott (2001)] or absorption refrigerator machines [Yang *et al* (2009), Ondryas *et al* (1991), Punwani (1999) and Kakarus *et al* (2004)]. In general, application of the mechanical air-cooling increases the net power but in the same time reduces the thermal efficiency. For example, Alhazmy *et al* (2004) showed that for a GT of pressure ratio 8 cooling the intake air from 50°C to 40°C increases the power by 3.85 % and reduces the thermal efficiency by 1.037%. Stewart and Patrick (2000) raised another disadvantage (for extensive air chilling) concerning ice formation either as ice crystals in the chilled air or as solidified layer on air compressors' entrance surfaces.

Recently, alternative cooling approaches have been investigated. Farzaneh-Gord and Deymi-Dashtebayaz (2009) proposed improving refinery gas turbines performance using the cooling capacity of refinerys' natural-gas pressure drop stations. Zaki *et al* 2007 suggested a reverse Brayton refrigeration cycle for cooling the air intake; they reported an increase in the output power up to 20%, but a 6% decrease in thermal efficiency. This approach was further extended by Jassim *et al* (2009) to include the exergy analysis and show that the second law analysis improvement has dropped to 14.66% due to the components irreversibilities. Khan *et al* (2008) analyzed a system in which the turbine exhaust gases are cooled and fed back to the compressor inlet with water harvested out of the combustion products. Erickson (2003, 2005) suggested using a combination of a waste heat driven absorption air cooling with water injection into the combustion air; the concept is named the *"power fogger cycle"*.

Thermal analyses of GT cooling are abundant in the literature, but few investigations considered the economics of the cooling process. A sound economic evaluation of implementing an air intake GT cooling system is quite involving. Such an evaluation should account for the variations in the ambient conditions (temperature and relative humidity) and the fluctuations in the fuel and electricity prices and interest rates. Therefore, the selection of a cooling technology (evaporative or refrigeration) and the sizing out of the equipment should not be based solely on the results of a thermal analysis but should include estimates of the cash flow. Gareta *et al* (2004) has developed a methodology for combined cycle GT that calculated the additional power gain for 12 months and the economic feasibility of the cooling method. From an economical point of view, they provided straight forward information that supported equipment sizing and selection. Chalker *et al* (2003) have studied the economical potential of using evaporative cooling for GTs in USA, while Hasnain (2002) examined the use of ice storage methods for GTs' air cooling in KSA. Yang *et al* (2009) presented an analytical method for evaluating a cooling technology of a combined cycle GT that included parameters such as the interest rate, payback period and the efficiency ratio for off-design conditions of both the GT and cooling system. Investigations of evaporative cooling and steam absorption machines showed that inlet fogging is superior in efficiency up to intake temperatures of 15-20°C, though it results in a smaller profit than inlet air chilling [Yang *et al* 2009].

In the present study, the performance of a cooling system that consists of a chilled water external loop coupled to the GT entrance is investigated. The analysis accounts for the changes in the thermodynamics parameters (applying the first and second law analysis) as well as the economic variables such as profitability, cash flow and interest rate. An objective of the present study is to assess the importance of using a coupled thermo-economics analysis in the selections of the cooling system and operation parameters. The developed algorithm is applied to an open cycle, HITACH MS-7001B plant in the hot weather of KSA (Latitude 24° 05″ N and longitude 38° E) by The result of this case study are presented and discussed.

2. GT-air cooling chiller energy analysis

Figure 1.a shows a schematic of a simple open GT "Brayton cycle" coupled to a refrigeration system. The power cycle consists of a compressor, combustion chamber and a turbine. It is presented by states 1-2-3-4 on the T-S diagram, Fig. 1.b. The cooling system consists of a refrigerant compressor, air cooled condenser, throttle valve and water cooled evaporator. The chilled water from the evaporator passes through a cooling coil mounted at the air compressor entrance, Fig. 1.a. The refrigerant cycle is presented on the T-S diagram, Figure 1.c, by states *a, b, c* and *d*. A fraction of the power produced by the turbine is used to power the refrigerant compressor and the chilled water pumps, as indicated by the dotted lines in Fig. 1.a. To investigate the performance of the coupled GT-cooling system the different involved cycles are analyzed in the following employing the first and second laws of thermodynamics.

2.1 Gas turbine cycle

As seen in Figures 1.a and 1.b, processes 1-2$_s$ and 3-4$_s$ are isentropic. Assuming the air as an ideal gas, the temperatures and pressures are related to the pressure ratio, PR, by:

$$\frac{T_{2s}}{T_1} = \frac{T_3}{T_{4s}} = \left[\frac{P_2}{P_1}\right]^{\frac{k-1}{k}} = PR^{\frac{k-1}{k}} \tag{1}$$

The net power output of a GT with mechanical cooling system as seen in Fig. 1.a is

$$\dot{W}_{net} = \dot{W}_t - (\dot{W}_{comp} + \dot{W}_{el,ch}) \tag{2}$$

The first term of the RHS is the power produced by the turbine due to expansion of hot gases;

$$\dot{W}_t = \dot{m}_t\, c_{pg}\eta_t \left(T_3 - T_{4s}\right). \tag{3}$$

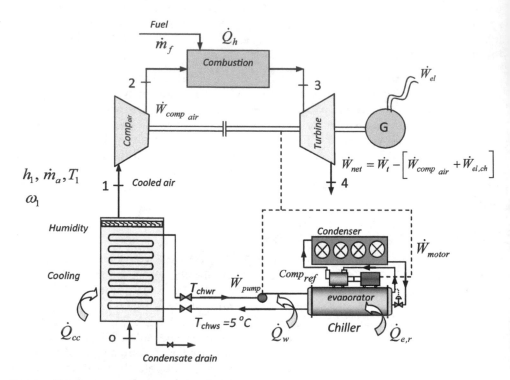

Fig. 1a. Simple open type gas turbine with a chilled air-cooling unit

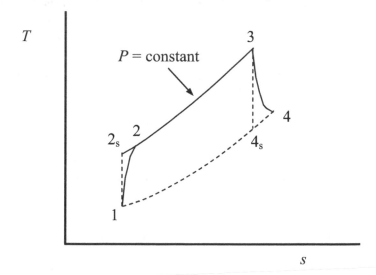

Fig. 1b. T-s diagram of an open type gas turbine cycle

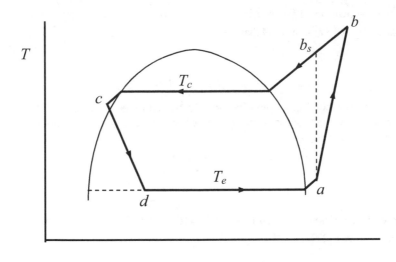

Fig. 1c. *T-s* diagram for a refrigeration machine

In Eq.3, $\dot{m}_t$ is the total gases mass flow rate from the combustion chamber; expressed in terms of the fuel air ratio $f = \dot{m}_f / \dot{m}_a$, and the air humidity ratio at the compressor intake ω_1, (kg$_w$/kg$_{dry\ air}$) (Fig. 1.a) as;

$$\dot{m}_t = \dot{m}_a + \dot{m}_v + \dot{m}_f = \dot{m}_a(1 + \omega_1 + f) \tag{4}$$

The compression power for humid air between states 1 and 2 is estimated from:

$$\dot{W}_{comp} = \dot{m}_a c_{pa}(T_2 - T_1) + \dot{m}_v(h_{v2} - h_{v1}) \tag{5}$$

where h_{v2} and h_{v1} are the enthalpies of saturated water vapor at the compressor exit and inlet states respectively, $\dot{m}_v$ is the mass of water vapor = $\dot{m}_a \omega_1$.

The last term in Eq. 2 ($\dot{W}_{el,ch}$) is the power consumed by the cooling unit for driving the refrigeration machine electric motor, pumps and auxiliaries.

The thermal efficiency of a GT coupled to an air cooling system is then;

$$\eta_{cy} = \frac{\dot{W}_t - (\dot{W}_{comp} + \dot{W}_{el,ch})}{\dot{Q}_h} \tag{6}$$

Substituting for T_{4s} and $\dot{m}_t$ from Equations (1) and (4) into Eq. (3) yields:

$$\dot{W}_t = \dot{m}_a(1 + \omega_1 + f)c_{pg}\eta_t T_3\left(1 - \frac{1}{PR^{\frac{k-1}{k}}}\right) \tag{7}$$

The turbine isentropic efficiency, η_t, can be estimated using the practical relation recommended by Alhazmy and Najjar (2004):

$$\eta_t = 1 - \left(0.03 + \frac{PR - 1}{180} \right) \tag{8}$$

Relating the compressor isentropic efficiency to the changes in temperature of the dry air and assuming that the compression of water vapor changes the enthalpy; the actual compressor power becomes;

$$\dot{W}_{comp_{air}} = \dot{m}_a \left[c_{pa} \frac{T_1}{\eta_c} \left(PR^{\frac{k-1}{k}} - 1 \right) + \omega_1 \left(h_{v2} - h_{v1} \right) \right] \tag{9}$$

The compression efficiency, η_c, can be evaluated using the following empirical relation, Alhazmy and Najjar (2004);

$$\eta_c = 1 - \left(0.04 + \frac{PR - 1}{150} \right) \tag{10}$$

The heat balance in the combustion chamber (Fig. 1.a) gives the heat rate supplied to the gas power cycle as:

$$\dot{Q}_h = \dot{m}_f NCV \eta_{comb} = \left(\dot{m}_a + \dot{m}_f \right) c_{pg} T_3 - \dot{m}_a c_{pa} T_2 + \dot{m}_v \left(h_{v3} - h_{v2} \right) \tag{11}$$

Introducing the fuel air ratio $f = \dot{m}_f / \dot{m}_a$ and substituting for T_2 in terms of T_1 into Eq.11 yields:

$$\dot{Q}_h = \dot{m}_a T_1 \left[(1+f) c_{pg} \frac{T_3}{T_1} - c_{pa} \left(\frac{PR^{\frac{k-1}{k}} - 1}{\eta_c} + 1 \right) + \frac{\omega_1}{T_1} (h_{v3} - h_{v2}) \right] \tag{12}$$

A simple expression for f is selected here, Alhazmy $et.al$ (2006) as:

$$f = \frac{c_{pg} (T_3 - 298) - c_{pa} (T_2 - 298) + \omega_1 (h_{v3} - h_{v2})}{NCV \eta_{comb} - c_{pg} (T_3 - 298)} \tag{13}$$

In equation 13, h_{v2} and h_{v3} are the enthalpies of water vapor at the combustion chamber inlet and exit states respectively and can be calculated from Eq. 14, Dossat (1997).

$$h_{v,j} = 2501.3 + 1.8723 \, T_j \quad j \textit{ refers to states 2 or 3} \tag{14}$$

The four terms of the gas turbine net power and efficiency in Eq. (6) ($\dot{W}_t$, $\dot{W}_{comp}$, $\dot{W}_{el,ch}$ and $\dot{Q}_h$) depend on the air temperature and relative humidity at the compressor inlet whose values are affected by the type and performance of the cooling system. The chillers' electric power, $\dot{W}_{el,ch}$, is calculated in the following account.

2.2 Refrigeration cooling system analysis

The chilled water from the refrigeration machine is the heat transport fluid to cool the intake air, Fig. 1.a. The chiller's total electrical power can be expressed as the sum of the electric motor power ($\dot{W}_{motor}$), the pumps ($\dot{W}_P$) and auxiliary power for fans and control units, ($\dot{W}_A$) as:

$$\dot{W}_{el,ch} = \dot{W}_{motor} + \dot{W}_P + \dot{W}_A \tag{15}$$

The auxiliary power is estimated as 10% of the compressor power, therefore, $\dot{W}_A = 0.1\,\dot{W}_{motor}$. The second term in Eq. 15 is the pumping power that is related to the chilled water flow rate and the pressure drop across the cooling coil, so that:

$$\dot{W}_P = \dot{m}_{cw}v_f\left(\Delta P\right)/\eta_{pump} \tag{16}$$

The minimum energy utilized by the refrigerant compressor is that for the isentropic compression process (a-b_s), Fig 1.c. The actual power includes losses due to mechanical transmission, inefficiency in the drive motor converting electrical to mechanical energy and the volumetric efficiency, Dossat (1997). The compressor electric motor work is related to the refrigerant enthalpy change as

$$\dot{W}_{motor} = \frac{\dot{m}_r\left(h_b - h_a\right)_r}{\eta_{eu}} \tag{17}$$

The subscript r indicates refrigerant and η_{eu} known as the energy use factor; $\eta_{eu} = \eta_m * \eta_{el} * \eta_{vo}$. The quantities on the right hand side are the compressor mechanical, electrical and volumetric efficiencies respectively. η_{eu} is usually determined by manufacturers and depends on the type of the compressor, the pressure ratio (P_b / P_a) and the motor power. For the present analysis η_{eu} is assumed 85%.

Cleland *et al* (2000) developed a semi-empirical form of Equation 17 to calculate the compressor's motor power usage in terms of the temperatures of the evaporator and condenser in the refrigeration cycle, T_e and T_c respectively as;

$$\dot{W}_{motor} = \frac{\dot{m}_r\left(h_a - h_d\right)_r}{\dfrac{T_e}{\left(T_c - T_e\right)}\left(1 - ax\right)^n \eta_{eu}} \tag{18}$$

In this equation, α is an empirical constant that depends on the type of refrigerant and x is the quality at state d, Fig 1.c. The empirical constant is 0.77 for R-22 and 0.69 for R-134a Cleland *et al* (2000). The constant n depends on the number of the compression stages; for a simple refrigeration cycle with a single stage compressor $n = 1$. The nominator of Eq. 18 is the evaporator capacity, $\dot{Q}_{e,r}$ and the first term of the denominator is the coefficient of performance of an ideal refrigeration cycle. Equations 2, 5 and 18 could be solved for the power usages by the different components of the coupled GT-refrigeration system to estimate the increase in the power output as function of the air intake conditions. Follows is a thermodynamics second law analysis to estimate the effect of irreversibilities on the power gain and efficiency.

3. Exergy analysis

In general, the expression for the exergy destruction, (Kotas 1995), is.

$$\dot{I} = T_o \left[\left(\dot{S}_{out} - \dot{S}_{in} \right) - \sum_{i=1}^{n} \frac{\dot{Q}_i}{T_i} \right] \ge 0 \tag{19}$$

and the exergy balance for any component of the coupled GT and refrigeration cooling cycle (Fig.1) is expressed as;

$$\dot{E}_{in} + \dot{E}^Q = \dot{E}_{out} + \dot{W} + \dot{I} \tag{20}$$

Various amounts of the exergy destruction terms due to irreversibility for each component in the gas turbine and the proposed air cooling system are given in final expressions, Table 1. Details of derivations can be found in Jassim, *et al* (2005 & 2009) and Khir *et.al* 2007.

Air Compressor 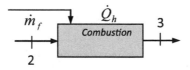
Air compressor process 1-2, Fig. 1-b
$$\dot{I}_{comp,air} = \dot{m}_a \left(1+\omega_1\right) T_o \left[c_{pa} \, \ell n \left(\frac{T_2}{T_1} \right) - R_a \, \ell n \left(\frac{P_2}{P_1} \right) \right] \tag{21}$$ $$\dot{W}_{eff,comp} = \dot{W}_{comp} + \dot{I}_{comp} \tag{22}$$
Combustion chamber 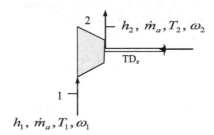
$$\dot{I}_{comb\,chamber} = \dot{m}_a T_o \left\{ \begin{array}{l} \left(1+f+\omega_1\right) \left[c_{pg} \, \ell n \left(\frac{T_3}{T_o} \right) - R_g \, \ell n \left(\frac{P_3}{P_o} \right) \right] - \\ \left(1+\omega_1\right) \left[c_{pa} \, \ell n \left(\frac{T_2}{T_o} \right) - R_a \, \ell n \left(\frac{P_2}{P_o} \right) \right] \end{array} \right\} + T_o \Delta S_o \tag{23}$$
$T_o \Delta S_o$ = rate of exergy loss in combustion or reaction = $\dot{m}_a \times f \times NCV \left(\varphi - 1 \right)$

Typical values of φ for some industrial fuels are given by Jassim *et al*, 2009, the effective heat to the combustion chamber

$$\dot{Q}_{eff,comb} = \dot{Q}_{comb} + \dot{I}_{comb} \tag{24}$$

Gas turbine

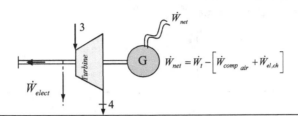

$$\dot{W}_{net} = \dot{W}_t - \left[\dot{W}_{comp\ air} + \dot{W}_{el,ch}\right]$$

$$\dot{I}_{gas\,turbine} = \dot{m}_a \left(1 + f + \omega_1\right) T_o \left[c_{pg}\,\ell n\left(\frac{T_4}{T_3}\right) - R_g\,\ell n\left(\frac{P_4}{P_3}\right)\right] \tag{25}$$

$$\dot{W}_{eff,t} = \dot{W}_t - \dot{I}_t \tag{26}$$

Chiller compressor

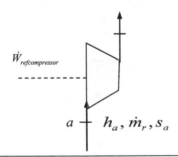

$$\dot{I}_{ref\ comp} = \dot{m}_r\,T_o\,(s_b - s_a) \tag{27}$$

Chiller Condenser

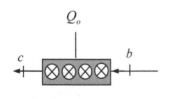

Refrigerant Condenser

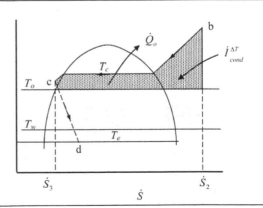

$$\dot{I}_{cond}^{\Delta T} = \dot{m}_r\, T_o \left[(s_c - s_b) + \frac{(h_b - h_c)}{T_o} \right] \tag{28}$$

The condenser flow is divided into three regions: superheated vapor region, two phase (saturation) region, and subcooled liquid region for which the exergy destruction due to flow pressure losses in each region are $\dot{I}_{cond,sup}^{\Delta P}$, $\dot{I}_{cond,sat}^{\Delta P}$ and $\dot{I}_{cond,sub}^{\Delta P}$. (Jassim *et al* 2005)

$$\dot{I}_{cond}^{\Delta P} = \dot{I}_{cond,sup}^{\Delta P} + \dot{I}_{cond,sat}^{\Delta P} + \dot{I}_{cond,sub}^{\Delta P} \tag{29}$$

$$\dot{I}_{cond} = \dot{I}_{cond}^{\Delta T} + \dot{I}_{cond}^{\Delta P} \tag{30}$$

Chiller cooling coil

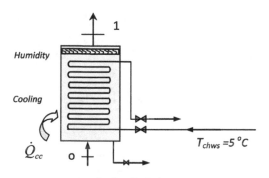

$$\dot{I}_{cooling\,coil} = \dot{m}_a \left(1 + \omega_1\right) T_o \left(s_o - s_1\right) + \dot{Q}_{out} \tag{31}$$

Expansion valve

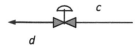

$$\dot{I}_{exp} = \dot{m}_r\, T_o \left[(s_d - s_c) \right] \tag{32}$$

Refrigerant evaporator

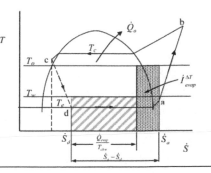

$$\dot{I}_{evap}^{\Delta T} = \dot{m}_r \, T_o \left[(s_a - s_d) - \frac{(h_a - h_d)}{T_{sw}} \right] \tag{33}$$

The refrigerant flow in the evaporator is divided into two regimes saturation(two phase) and superheated regions. The two phase (saturation) region, and superheated vapor region for which the exergy destruction due to flow pressure losses in each region are $\dot{I}_{evap,sat}^{\Delta P}$, $\dot{I}_{evap,sup}^{\Delta P}$ see Khir *et al* 2007. The exergy destruction rate is the sum of the thermal and pressure loss terms for both regimes (Eqs. 34 and 35) as,

$$\dot{I}_{evap} = \dot{I}_{evap}^{\Delta T} + \dot{I}_{evap}^{\Delta P} \tag{34}$$

$$\dot{I}_{evap}^{\Delta P} = \dot{I}_{evap,sat}^{\Delta P} + \dot{I}_{evap,sup}^{\Delta P} \tag{35}$$

Table 1. Exergy destruction terms for the individual components of the GT and coupled cooling chilled water unit, see Figs 1.a-1.c

4. Economics analysis

The increase in the power output due to intake air cooling will add to the revenue of the GT plant but will partially offset by the increase of the annual payments associated with the installation, personnel and utility expenditures for the operation of that system. For a cooling unit that includes a water chiller, the increase in expenses include the capital installments for the chiller, $\left(C_{ch}^c \right)$, and cooling coil, $\left(C_{cc}^c \right)$. The annual operation expenses is a function of the operation period, t_{op}, and the electricity rate. If the chiller consumes electrical power $\dot{W}_{el,ch}$ and the electricity rate is C_{el} ($/kWh) then the total annual expenses can be expressed as:

$$C_{total} = a^c \left[C_{ch}^c + C_{cc}^c \right] + \int_0^{t_{op}} C_{el} \, \dot{W}_{el,ch} \, dt \quad (\$/y) \tag{36}$$

In equation 36, the capital recovery factor $a^c = \dfrac{i(1+i)^n}{(1+i)^n - 1}$, which when multiplied by the total investment gives the annual payment necessary to payback the investment after a specified period (n).

The chiller's purchase cost may be estimated from venders data or mechanical equipment cost index; this cost is related to the chiller's capacity, $\dot{Q}_{e,r}$ (kW). For a particular chiller size and method of construction and installation; the capital cost is usually given by manufacturers in the following form;

$$C_{ch}^c = \alpha_{ch} \, \dot{Q}_{e,r} \tag{37}$$

For simplicity, the maintenance expenses are assumed as a fraction, α_m, of the chiller capital cost, therefore, the total chiller cost is expressed as;

$$C_{ch}^c = \alpha_{ch} \left(1 + \alpha_m\right) \dot{Q}_{e,r} \quad (\$) \tag{38}$$

Similarly, the capital cost of a particular cooling coil is given by manufacturers in terms of the cooling capacity that is directly proportional to the total heat transfer surface area (A_{cc}, m²) Kotas (1995) as;

$$C_{cc}^c = \beta_{cc} \left(A_{cc}\right)^m \quad (\$) \tag{39}$$

In equation 39, β_{cc} and m depend on the type of the cooling coil and material. For the present study and local Saudi market, $\beta_{cc} = 30000$ and $m = 0.582$ are recommended (York Co consultation, 2009). Substituting equations 38 and 39 into Eq. 36, assuming for simplicity that the chiller power is an average constant value and constant electricity rate over the operation period, the annual total expenses for the cooling system become;

$$C_{total} = a^c \left[\alpha_{ch} \left(1 + \alpha_m\right) \dot{Q}_{e,r} + \beta_{cc} \left(A_{cc}\right)^m \right] + t_{op} C_{el} \dot{W}_{el,ch} \quad (\$/y) \tag{40}$$

In Eq. 40 the heat transfer area A_{cc} is the parameter used to evaluate the cost of the cooling coil. Energy balance on both the cooling coil and the refrigerant evaporator, taking into account the effectiveness factors for the evaporator, $\varepsilon_{eff,er}$, and the cooling coil, $\varepsilon_{eff,cc}$, gives

$$A_{cc} = \frac{\dot{Q}_{cc}}{U \, \Delta T_m F \varepsilon_{eff,cc}} = \frac{\dot{Q}_{e,r} \, \varepsilon_{eff,er}}{U \, \Delta T_m F} \tag{41}$$

Where, U is the overall heat transfer coefficient for chilled water-air tube bank heat exchanger. Gareta, et al (2004) suggested a moderate value of 64 W/m² K and 0.98 for the correction factor F.

Figure 2, illustrates the temperature variations in the combined refrigerant, water chiller and air cooling system. the mean temperature difference for the cooling coil (air and chilled water fluids) is;

$$\Delta T_m = \frac{\left(T_o - T_{chwr}\right) - \left(T_1 - T_{chws}\right)}{\ell n \left(\left(T_o - T_{chwr}\right) / \left(T_1 - T_{chws}\right)\right)} \tag{42}$$

Equations 39 and 41 give the cooling coil cost as,

$$C_{cc}^c = \beta_{cc} \left(\frac{\dot{Q}_{cc}}{U \, \Delta T_m F}\right)^m \tag{43}$$

where, $\dot{Q}_{cc}$ is the thermal capacity of the cooling coil. The atmospheric air enters at T_o and ω_o and leaves the cooling coil to enter the air compressor intake at T_1 and ω_1, Fig.1.a. Both T_1 and ω_1 depend on the chilled water supply temperature (T_{chws}) and mass flow rate, $\dot{m}_{cw}$. When the outer surface temperature of the cooling coil falls below the dew point (corresponding to the partial pressure of the water vapor) the water vapor condensates and leaves the air stream. This process may be treated as a cooling-dehumidification process as illustrated in Figure 3. Steady state heat balance of the cooling coil gives;

$$\dot{Q}_{cc} = \dot{m}_a\left(h_o - h_1\right) - \dot{m}_w h_w = \dot{m}_{cw} c_w \,\varepsilon_{eff,cc}\left(T_{chwr} - T_{chws}\right) \qquad (44)$$

where, $\dot{m}_{cw}$ is the chilled water mass flow rate and $\dot{m}_w$ is the rate of water extraction from the air, $\dot{m}_w = \dot{m}_a\left(\omega_o - \omega_1\right)$. The second term in equation 44 is usually a small term when compared to the first and can be neglected, McQuiston $et\ al$ (2005).

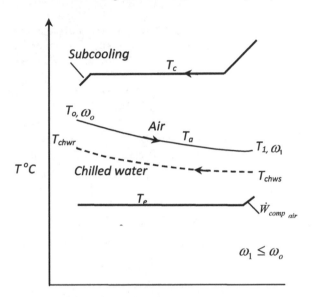

Fig. 2. Temperature levels for the three working fluids, not to scale

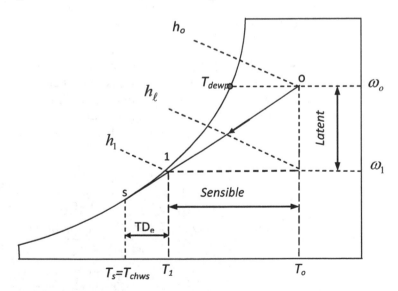

Fig. 3. Moist air cooling process before GT compressor intake

In equation 44 the enthalpy and temperature of the air leaving the cooling coil (h_1 and T_1) may be calculated from;

$$h_1 = h_o - CF(h_o - h_s) \tag{45}$$

$$T_1 = T_o - CF(T_o - T_s) \tag{46}$$

The contact factor CF is defined as the ratio between the actual air temperature drop to the maximum, at which the air theatrically leaves at the coil surface temperature $T_s = T_{chws}$ and 100% relative humidity. Substituting for h_1 from Eq. 45 into Eq. 44 and use Eq. 41 gives;

$$\dot{Q}_{e,r} = \frac{\dot{m}_a \left[CF(h_o - h_{chws}) - (\omega_o - \omega_1)h_w \right]}{\varepsilon_{eff,er}\,\varepsilon_{eff,cc}} \tag{47}$$

Equations 40 through 47 give the chiller and cooling coil annual expenses in terms of the air mass flow rate and properties. The total annual cost function is derived from Eq. 40 as follows.

4.1 Annual cost function

Combining equations 40 and 41, substituting for the cooling coil surface area, pump and auxiliary power gives the total annual cost in terms of the evaporator capacity $\dot{Q}_{er}$, as,

$$C_{total} = \left\{ \begin{array}{l} a^c \left[\alpha_{ch}(1+\alpha_m)\dot{Q}_{er} + \beta_{cc}\left(\dfrac{\dot{Q}_{er}\,\varepsilon_{eff,er}\,\varepsilon_{eff,cc}}{U\,\Delta T_m F} \right)^m \right] + \\[3mm] t_{op}\dot{Q}_{er}C_{el}\left[\left(\dfrac{1.1(T_c - T_e)}{T_e(1-\alpha x)^n\,\eta_{eu}} \right) + \left(\dfrac{\varepsilon_{eff,er}\,v_f(\Delta P)}{c_{p,w}\,\Delta T_{ch,w}\eta_{pump}} \right) \right] \end{array} \right\} \tag{48}$$

The first term in Eq. 48 is the annual fixed charges of the refrigeration machine and the surface air cooling coil, while the second term is the operation expenses that depend mainly on the electricity rate. If the water pump's power is considered small compared to the compressor power, the second term of the operation charges can be dropped. If the evaporator capacity $\dot{Q}_{er}$ is replaced by the expression in Eq. 47, the cost function, in terms of the primary parameters, becomes;

$$C_{total} = \left[\frac{\dot{m}_a\left[CF(h_o - h_{chws}) - (\omega_o - \omega_1)h_w \right]}{\varepsilon_{eff,er}\,\varepsilon_{eff,cc}} \right] \left\{ a^c \left[\begin{array}{l} \alpha_{ch}(1+\alpha_m) + \beta_{cc}\left(\dfrac{\varepsilon_{eff,er}\,\varepsilon_{eff,cc}}{U\,\Delta T_m F} \right)^m \times \\[3mm] \left(\dfrac{\dot{m}_a\left[CF(h_o - h_{chws}) - (\omega_o - \omega_1)h_w \right]}{\varepsilon_{eff,er}\,\varepsilon_{eff,cc}} \right)^{m-1} \\[3mm] +t_{op}C_{el}\left[\left(\dfrac{1.1(T_c - T_e)}{(T_e)(1-ax)^n\,\eta_{eu}} \right) + \left(\dfrac{\varepsilon_{eff,er}\,v_f(\Delta P)}{c_{p,w}\,\Delta T_{ch,w}\eta_p} \right) \right] \end{array} \right] \right\} \tag{49}$$

5. Evaluation criteria of GT-cooling system

In order to evaluate the feasibility of a cooling system coupled to a GT plant, the performance of the plant is examined with and without the cooling system. In the present study it is recommended to consider the results of the three procedures (energy, exergy and economics analysis).

5.1 First law efficiency

In general, the net power output of a complete system is given in Eq. 2 in terms of $\dot{W}_t, \dot{W}_{comp}$ and $\dot{W}_{el,ch}$. The three terms are functions of the air properties at the compressor intake (T_1 and ω_1), which in turn depend on the performance of the cooling system. The present analysis considers the "*power gain ratio*" (*PGR*), a broad term suggested by AlHazmy *et al* (2006) that takes into account the operation parameters of the GT and the associated cooling system:

$$PGR = \frac{\dot{W}_{net,with cooling} - \dot{W}_{net,without\ cooling}}{\dot{W}_{net,without\ cooling}} \times 100\% \qquad (50)$$

For a stand-alone GT, *PGR* = 0. Thus, the *PGR* gives the percentage enhancement in power generation by the coupled system. The thermal efficiency of the system is an important parameter to describe the input-output relationship. The *thermal efficiency change factor* (*TEC*) proposed in AlHazmy *et al* (2006) is defined as

$$TEC = \frac{\eta_{cy,with cooling} - \eta_{cy,without\ cooling}}{\eta_{cy,without\ cooling}} \times 100\% \qquad (51)$$

5.2 Exrgetic efficiency

Exergetic efficiency is a performance criterion for which the output is expressible in terms of exergy. Defining the exergetic efficiency η_{ex}, as a ratio of total rate of exergy output $\left(\dot{E}_{out}\right)$ to total rate of exergy input $\left(\dot{E}_{in}\right)$ as;

$$\eta_{ex} = \frac{\dot{E}_{out}}{\dot{E}_{in}} \qquad (52)$$

The exergy balance for the gas turbine and the water chiller system, using the effective work and heat terms in Table 1, can be expressed in the following forms,

$$\dot{E}_{out} = \dot{W}_{eff,t} - \dot{W}_{eff,comp} - \dot{W}_{eff,Chiller} \qquad (53)$$

and

$$\dot{E}_{in} = \dot{Q}_{eff,comb} - \dot{Q}_{eff,cc} \qquad (54)$$

In analogy with the energy efficiency the exergetic efficiency for a GT-refrigeration unit is:

$$\eta_{ex,c} = \frac{\dot{W}_{eff,t} - \dot{W}_{eff,comp} - \dot{W}_{eff,chiller}}{\dot{Q}_{eff,comb} - \dot{Q}_{eff,cc}} \tag{55}$$

For the present analysis let us define dimensionless terms as the *exergetic power gain ratio* (*PGR_ex*) and *exergetic thermal efficiency change* (*TEC_ex*):

$$PGR_{ex} = \frac{\left(\dot{E}_{out}\right)_{withcooling} - \left(\dot{E}_{out}\right)_{without cooling}}{\left(\dot{E}_{out}\right)_{without cooling}} \times 100\% \tag{56}$$

and

$$TEC_{ex} = \frac{\eta_{ex,c} - \eta_{ex,nc}}{\eta_{ex,nc}} \times 100\% \tag{57}$$

Equations 50, 51, 56 and 57 can be easily employed to appraise the changes in the system performance, but they are not sufficient for a complete evaluation of the cooling method, the economics assessement of installing a cooling system follows.

5.3 System profitability

To investigate the economic feasibility of retrofitting a gas turbine plant with an intake cooling system, the total cost of the cooling system is determined (Eq. 32 or Eq. 33). The increase in the *annual* income cash flow from selling the additional electricity generation is also calculated. The annual exported energy by the coupled power plant system is;

$$E\,(kWh) = \int_0^{t_{op}} \dot{W}_{net} dt \tag{58}$$

If the gas turbine's annual electricity generation without the cooling system is $E_{without\ cooling}$ and the cooling system increases the power generation to $E_{with\ cooling}$, then the net increase in revenue due to the addition of the cooling system is:

$$Net\ revenue = (E_{with\ cooling} - E_{without\ cooling})C_{els} \tag{59}$$

The profitability due to the coupled power plant system is defined as the increase in revenues due to the increase in electricity generation after deducting the expenses for installing and operating the cooling system as:

$$rofitability = (E_{with\ cooling} - E_{without\ cooling})C_{els} - C_{total} \tag{60}$$

The first term in Eq. 60 gives the increase in revenue and the second term gives the annual expenses of the cooling system. The profitability could be either positive, which means an economical incentive for adding the cooling system, or negative, meaning that there is no economical advantage, despite the increase in the electricity generation of the plant.

For more accurate evaluation the irreversibility of the different components are taken into consideration and an effective revenue *(Revenue)_eff* is defined by;

$$Revenue_{eff} = \int_{0}^{t_{op}} \left(\left(\dot{E}_{out} \right)_{with\,cooling} - \left(\dot{E}_{out} \right)_{without\,cooling} \right) C_{els} \, dt \qquad (61)$$

6. Results and discussion

The performance of the GT with water chiller cooler and its economical feasibility are investigated. The selected site is the Industrial City of Yanbu (Latitude 24° 05' N and longitude 38° E) where a HITACH FS-7001B model GT plant is already connected to the main electric grid. Table 2 lists the main specs of the selected GT plant. The water chiller capacity is selected on basis of the maximum annual ambient temperature at the site. On August 18th, 2008, the dry bulb temperature (DBT) reached 50°C at 14:00 O'clock and the relative humidity was 84% at dawn time. The recorded hourly variations in the DBT (T_o) and RH_o are shown in Figure 4 and the values are listed in Table 2. Eq. 47 gives the evaporator capacity of the water chiller (Ton Refrigeration) as function of the DBT and RH. Figure 5 shows that if the chiller is selected based on the maximum DBT = 50°C and RH = 18%, (the data at 14: O'clock), its capacity would be 2200 Ton. Another option is to select the chiller capacity based on the maximum RH_o (RH_o = 0.83 and T_o = 28.5°C, 5:00 data), which gives 3500 Ton. It is more accurate, however, to determine the chiller capacity for the available climatic data of the selected day and determine the maximum required capacity, as seen in Fig. 6; for the weather conditions at Yanbu City, a chiller capacity of 4200 Ton is selected it is the largest chiller capacity $\left(\dot{Q}_{e,r} \right)$ to handle the worst scenario as shown in Fig. 6.

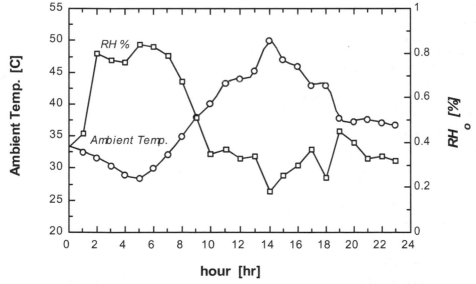

Fig. 4. Ambient temperature variation and RH for 18th of August 2008 of Yanbu Industrial City

Equations 45 and 46 are employed to give the air properties leaving the cooling coil, assuming 0.5 contact factor and a chilled water supply temperature of 5°C. All thermophysical properties are determined to the accuracy of the Engineering Equation Solver (EES)

software [Klein and Alvarado 2004]. The result show that the cooling system decrease the intake air temperature from T_o to T_1 and increases the relative humidity to RH_1 (Table 3).

Parameter	Range
Ambient air, Fig. 4	
Ambient air temperature, T_o	28 − 50 °C
Ambient air relative humidity, RH_o	18% → 84%
Gas Turbine, Model HITACH-FS-7001B	
Pressure ratio, P_2/P_1	10
Net power, ISO	52.4 MW
Site power	37 MW
Turbine inlet temperature T_3	1273.15 K
Volumetric air flow rate	250 m³s⁻¹at NPT
Fuel net calorific value, NCV	46000 kJ kg⁻¹
Turbine efficiency, η_t	0.88
Air Compressor efficiency η_c	0.82
Combustion efficiency η_{comb}	0.85
Generator	
Electrical efficiency	95%
Mechanical efficiency	90%
Water Chiller	
Refrigerant	R22
Evaporating temperature, T_e	$T_{chws} - TD_e$ °C
Superheat	10 K
Condensing temperature, T_c	T_o + TD_c K
Condenser design temperature difference TD_c	10 K
Evaporator design temperature difference TDe	6 K
Subcooling	3 K
Chilled water supply temperature, T_{chws}	5 °C
Chiller evaporator effectiveness, $\varepsilon_{eff,er}$	85%
Chiller compressor energy use efficiency, η_{eu}	85%
α_{ch}	172 $/kW
Cooling Coil	
Cooling coil effectiveness $\varepsilon_{eff,cc}$	85%
Contact Factor, CF	50%
Economics analysis	
Interest rate i	10%
Period of repayment (Payback period), n	3 years
The maintenance cost, α_m	10% of C_{ch}^c
Electricity rate, C_{el} (Eqs. 33&34)	0.07 $/kWh
Cost of selling excess electricity, C_{els} (Eqs. 40&41)	0.07-0.15 $/kWh
Hours of operation per year, t_{op}	7240 h/y

Table 2. Range of parameters for the present analysis

Hour	$T_o °C$	RH_o	$T_1 °C$	RH_1	Hour	$T_o °C$	RH_o	$T_1 °C$	RH_1
0	33.4	0.38	19.2	0.64	12	44.0	0.33	24.5	0.64
1	32.6	0.44	18.8	0.70	13	45.2	0.34	25.1	0.66
2	31.7	0.8	18.35	0.99	14	50.0	0.18	27.5	0.43
3	30.5	0.77	17.75	0.98	15	47.0	0.25	26.0	0.53
4	29.0	0.76	17.0	0.99	16	45.9	0.30	25.45	0.61
5	28.5	0.84	16.75	0.97	17	43.0	0.37	24.0	0.69
6	30.0	0.83	17.5	0.99	18	43.0	0.24	24.0	0.50
7	32.2	0.79	18.6	0.96	19	37.9	0.45	21.45	0.76
8	35.1	0.67	20.05	0.99	20	37.4	0.40	21.2	0.69
9	38.0	0.51	21.5	0.84	21	37.6	0.33	21.3	0.60
10	40.2	0.35	22.6	0.64	22	37.1	0.34	21.05	0.61
11	43.3	0.37	24.15	0.69	23	36.8	0.32	20.90	0.58

Table 3. The ambient conditions and the cooling coil outlet temperature and humidity during 18th August 2008 operation

Solution of Equations 50 and 51, using the data in Table 3, gives the daily variation in *PGR* and *TEC*, Figure 7. There is certainly a potential benefit of adding the cooling system where there is an increase in the power output all the time, the calculated average for the design day is 12.25 %. The *PGR* follows the same pattern of the ambient temperature; the increase in power of the GT plant reaches a maximum of 15.46 %, with a little change in the plant thermal efficiency. The practical illustrative application indicates that a maximum decrease in the thermal efficiency change of only 0.391 % occurs at 13:00 PM when the air temperature is 45.2°C, and 34% RH.

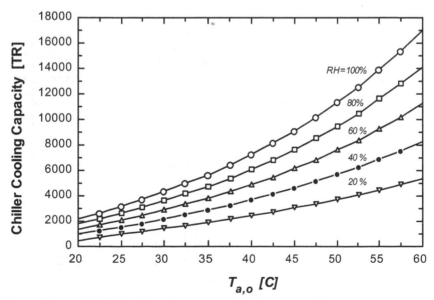

Fig. 5. Dependence of chiller cooling capacity on the climatic conditions

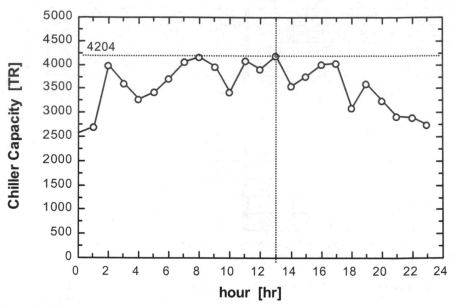

Fig. 6. Chiller capacity variation with the climatic conditions of the selected design day

On basis of the second law analysis the exergetic power gain ratio PGR_{ex} is still positive meaning that there is increase in output power but at a reduced value than that of the energy analysis.

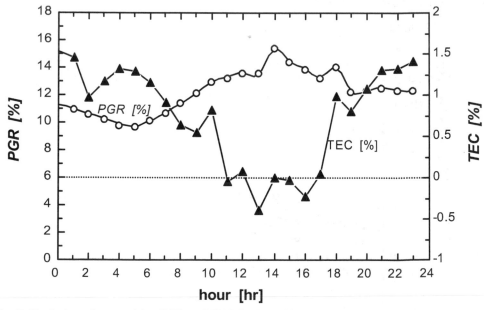

Fig. 7. Variation of gas turbine PGR and TEC during 18th August operation

Figure 8. shows that the power increase for the worst day of the year that varies between 7% to 10.4% (average 8.5%) and the thermal efficiency drops by a maximum of 6 %. These result indicate the importance of the second law analysis.

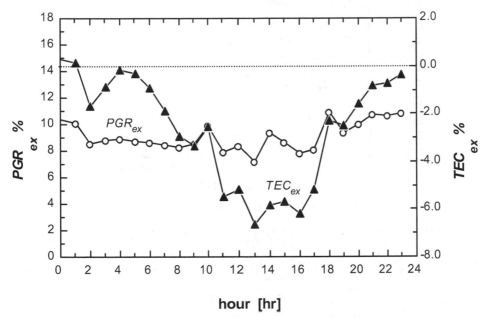

hour [hr]

Fig. 8. Variation of gas turbine exergetic PGR_{ex} and TEC_{ex} during 18th August operation

Based on the daily variation of the ambient conditions on August 18th, assuming different values for selling the electricity (C_{els}), Eq. 59 gives the hourly revenues needed to payback the investment after a specified operation period (selected by 3 years). The different terms in Equations 49 and 59 are calculated and presented in Figure 9. The effect of the climate changes is quite obvious on both the total expenses (Fig. 9) and the GT net power output (Fig. 7). The variations in C_{total} are due to the changes in $\dot{Q}_{ev}$ in Eq. 49 that depends on (T_o, T_1, ω_o and ω_1). The revenue from selling additional electricity is also presented in the same figure, which shows clearly the potential of adding the cooling system. Figure 9 indicates that selling the electricity to the consumers at the same base price ($C_{els} = C_{el}$ = 0.07 $/kWh) makes the cooling system barley profitable. The profit increases directly with the cost of selling the electricity. This result is interesting and encourages the utilities to consider a time-of-use tariff during the high demand periods. The profitability of the system, being the difference between the revenues and the total cost, is appreciable when the selling rate of the excess electricity generation is higher than the base rate of 0.07 $/kWh.

Economy calculations for one year with 7240 operation hours and for different electricity selling rates are summarized in Table 4. The values show that there is always a net positive profit starting after the payback period for different energy selling prices. During the first 3 years of the cooling system life, there is a net profit when the electricity selling rate increases to 0.15 $/kWh, nearly double the base tariff.

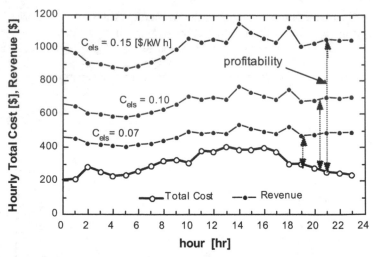

Fig. 9. Variation of hourly total cost and excess revenue at different electricity selling rate

Electricity selling rate C_{els}	Annuity-for Chiller, coil and maintenance	Annual operating cost	Annual net profit for the first 3 years	Annual net profit for the fourth year
$/kWh	$/y	$/y	$/y	$/y
0.07	1,154,780	1,835,038	-1,013,600	+141180
0.1	1,154,780	1,835,038	-166,821	+ 987,962
0.15	1,154,780	1,835,038	1,244,978	+ 2,399,758

Table 4. Annual net profits out of retrofitting a cooling system to a GT, HITACHI FS-7001B at Yanbu for different product tariff and 3 years payback period

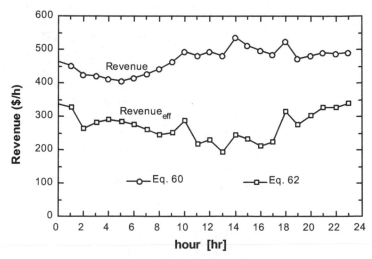

Fig. 10. Effect of irreversibility on the revenue, C_{els} = 0.07 $/kWh

Figure 10 shows the effect of irreversibilities on the economic feasibility of using an air cooling system for the selected case. The effective revenue Eq. 61 $\left(\text{Revenue}_{eff}\right)$ that can be accumulated from selling the net power output is reduced by 41.8% as a result of irreversibilities. The major contribution comes from the water chiller, where the irreversibility is the highest.

7. Conclusions

There are various methods to improve the performance of gas turbine power plants operating under hot ambient temperatures far from the ISO standards. One proven approach is to reduce the compressor intake temperature by installing an external cooling system. In this paper, a simulation model that consists of thermal analysis of a GT and coupled to a refrigeration cooler, exergy analysis and economics evaluation is developed. The performed analysis is based on coupling the thermodynamics parameters of the GT and cooler unit with the other variables as the interest rate, life time, increased revenue and profitability in a single cost function. The augmentation of the GT plant performance is characterized using the power gain ratio (PGR) and the thermal efficiency change term (TEC).

The developed model is applied to a GT power plant (HITACHI FS-7001B) in the city of Yanbu (20° 05″ N latitude and 38° E longitude) KSA, where the maximum DBT has reached 50°C on August 18th, 2008 The recorded climate conditions on that day are selected for sizing out the chiller and cooling coil capacities. The performance analysis of the GT shows that the intake air temperature decreases by 12 to 22 K, while the PGR increases to a maximum of 15.46%. The average increase in the plant power output power is 12.25%, with insignificant change in plant thermal efficiency. The second law analysis show that the exergetic power gain ratio drops to an average of 8.5% with 6% maximum decrease in thermal efficiency.

In the present study, the profitability resulting from cooling the intake air is calculated for electricity rates between 0.07 and 0.15 $/kWh and a payback period of 3 years. Cash flow analysis of the GT power plant in the city of Yanbu shows a potential for increasing the output power of the plant and increased revenues.

8. Nomenclatures

A_{cc}	Cooling coil heat transfer area, m²
C_{cc}^{c}	capital cost of cooling coil ($)
C_{ch}^{c}	capital cost of chiller ($)
C_{el}	unit cost of electricity, $/kWh
c_p	specific heat of gases, kJ/kg K
CF	contact factor
E	energy kWh
EES	engineering Equation Solver
h_v	specific enthalpy of water vapor in the air, kJ/kg
i	interest rate on capital

$\dot{I}$	exergy destruction, kW
k	specific heats ratio.
$\dot{m}$	mass flow rate, kg s^{-1}
$\dot{m}_a$	air mass flow rate, kg/s
$\dot{m}_{cw}$	chilled water mass flow rate, kg/s
$\dot{m}_r$	refrigerant mass flow rate, kg/s
$\dot{m}_w$	condensate water rate, kg/s
NCV	net calorific value, kJ kg^{-1}
P	pressure, kPa
PGR	power gain ratio
P_o	atmospheric pressure, kPa
PR	pressure ratio = P_2/P_1
$\dot{Q}_h$	heat rate, kW
$\dot{Q}_{e,r}$	chiller evaporator cooling capacity, kW
$\dot{Q}_{cc}$	cooling coil thermal capacity, kW
$\dot{S}$	entropy, kJ/K
t	time, s
T	Temperature, K
TEC	thermal efficiency change factor
U	overall heat transfer coefficient, kW/m^{2}K
x	quality.
$\dot{W}$	power, kW

Greek symbols

η	efficiency
ε_{eff}	effectiveness, according to subscripts
ω	specific humidity (also, humidity ratio),according to subscripts, kg/kg$_{dry\ air}$

Subscripts

a	dry air
c	with cooling
cc	cooling coil
ch	chiller
$comb$	combustion
$comp$	compressor
eff	effective
el	electricity
f	fuel
g	gas
nc	no cooling
o	ambient
t	turbine
v	vapor

9. References

Alhazmy MM, Jassim RK, Zaki GM. 2006. Performance enhancement of gas turbines by inlet air-cooling in hot and humid climates. *International Journal of Energy Research* 30:777-797

Alhazmy MM. and Najjar YS. 2004. Augmentation of gas turbine performance using air coolers, App. Thermal Engineering. 24: 415-429.

Ameri M, Nabati H. Keshtgar A. 2004. Gas turbine power augmentation using fog inlet cooling system. Proceedings ESDA04 7th Biennial conf. engineering systems design and analysis, Manchester UK. paper ESDA 2004-58101.

Ameri M, Shahbazian HR, Nabizadeh M. 2007. Comparison of evaporative inlet air cooling systems to enhance the gas turbine generated power. Int. J. Energy Res. 31: 483-503

Chaker M, Meher-Homji CB, Mee III, Nicholson A. 2003. Inlet fogging of gas turbine engines detailed climatic analysis of gas turbine evaporation cooling potential in the USA. *J Eng Gas Turbine Power* 125(1):300–309.

Chaker M., Meher-Homji, CB, Mee M. 2002. Inlet fogging of gas turbine engines - Part B: Fog droplet sizing analysis, nozzle types, measurement and testing, ASME Proceedings of Turbo Expo 2002, 4:429-442.

Chaker, M, Meher-Homji CB, Mee M. 2002. Inlet fogging of gas turbine engines - part c: fog behavior in inlet ducts, cfd analysis and wind tunnel experiments, ASME Proceedings of Turbo Expo 2002, Vol. 4:443-455.

Cleland AJ, Cleland DJ, White SD. 2000. *Cost-Effective Refrigeration,* Short course notes, Institute of Technology and Engineering, *Massey University, New Zealand*

Cortes CPE, Williams D. 2003. Gas turbine inlet cooling techniques: An overview of current technology. Proceedings Power GEN 2003, Las Vegas Nevada Dec. 9-11.

Dossat RJ. 1997. *Principles of Refrigeration,* John Wiley and Sons, NY.

Elliot J. 2001. Chilled air takes weather out of equation, *Diesel and gas turbine world wide,* Oct: 49-96.

Erickson DC. 2003. Aqua absorption turbine inlet cooling, *Proceedings of IMEC 03, ASME International Mechanical Engineering Congress & Exposition,* Nov. 16-21, Washington DC

Erickson DC. 2005. Power fogger cycle. *ASHRAE Transactions* 111, part 2:551-554.

Farzaneh-Gord M and Deymi-Dashtebayaz M. 2009. A new approach for enhancing performance of a gas turbine (case study: Khangiran refinery). *Applied Energy* 86: 2750–2759

Gajjar H, Chaker M. 2003. Inlet fogging for a 655 MW combined cycle power plant-design, implementation and operating experience, *ASME Proceedings of Turbo Expo 2003,* Vol. 2:853-860.

Gareta R, Romeo LM, Gil A. 2004. Methodology for the economic evaluation of gas turbine air cooling systems in combined cycle applications *Energy* 29:1805–1818.

Hameed Z, Personal communication, Salim York Co. Saudi Arabia

Hasnain SM, Alawaji SH, Al-Ibrahim AM, Smiai MS. 2000. Prospects of cool thermal storage utilization in Saudi Arabia. *Energy Conversion & Management* 41: 1829-1839.

Homji-Meher, BC, Mee T, Thomas R. 2002. Inlet fogging of gas turbine engines, part B: Droplet sizing analysis nozzle types, measurement and testing, *Proceedings of the ASME Turbo Expo 2002,* Amsterdam, Netherlands, June 2002 paper No: GT-30563.

Jassim RK, Zaki GM,. Alhazmy MM. 2009, Energy and Exergy Analysis of Reverse Brayton
    Refrigerator for Gas Turbine Power Boosting, Int. Journal of Exergy 6 (2): 143-165.
Jassim, RK, Khir T, Habeebullah BA and Zaki, GM. 2005. Exergoeconomic Optimization of
    the Geometry of Continuous Fins on an Array of Tubes of a Refrigeration Air
    Cooled Condenser, Int. Journal of Exergy. 2 (2): 146-171.
Jonsson M, Yan J. 2005. Humidified gas turbines- A review of proposed and implemented
    cycles, Energy 30: 1013-1078.
Kakarus E, Doukelis A, Karellas S. 2004. Compressor intake air cooling in gas turbine plants.
    Energy 29:2347-2358.
Khan JR, Lear WE, Sherif SA, Crittenden JF. 2008. Performance of a novel combined cooling
    and power gas turbine with water harvesting. ASME Journal of Engineering for Gas
    Turbines and Power 130 (4).
Khir T, Jassim RK, Zaki, GM. 2007. Application of Exergoeconomic Techniques to the
    optimization of a refrigeration evaporator coil with continuous fins. Trans, ASME,
    Journal of Energy Resources Technology 129 (3): 266-277.
Klein KA, Alvarado FL. 2004. EES-Engineering Equation Solver, Version 6.648 ND, F-Chart
    Software, Middleton, WI.
Kotas TJ. 1995. The exergy method of thermal plant analysis, Krieger, Malabar, Florida, USA.
McQuiston FC, Parker JD, Spilter JD. 2005. Heating, Ventilating and Air conditioning:
    Design and analysis, 6th edition, John Wily, NY.
Ondrays IS, Wilson DA, Kawamoto N, Haub GL. 1991. Options in gas turbine power
    augmentation using inlet air chilling. Eng. Gas Turbine and Power. 113: 203-211.
Punwani D, Pierson T, Sanchez C, Ryan W. 1999. Combustion turbine inlet air cooling using
    absorption chillers some technical and economical analysis and case summaries
    ASHRAE Annual Meeting, Seattle, Washington, June, 99.
Sanaye S, Tahani M. 2010. Analysis of gas turbine operating parameters with inlet fogging
    and wet compression processes, Applied thermal engineering, 30: 234-244.
Stewart W, Patrick A. 2000. Air temperature depression and potential icing at the inlet of
    stationary combustion turbines. ASHRAE Transactions 106, pt 2:318-327.
Tillman TC, Blacklund DW, Penton JD. 2005. Analyzing the potential for condensate carry-
    over from a gas cooling turbine inlet cooling coil, ASHRAE Transactions 111 (Part 2)
    DE-05-6-3: 555-563.
Wang T, Li X and Pinniti V. 2009. Simulation of mist transport for gas turbine inlet air-
    cooling. ASME Int. Mec. Eng congress, November 13-19. Anaheim, Ca, USA,
Yang C, Yang Z, Cai R. 2009. Analytical method for evaluation of gas turbine inlet air
    cooling in combined cycle power plant, Applied Energy 86:848–856
Zaki GM, Jassim RK, Alhazmy MM. 2007. Brayton refrigeration Cycle for gas turbine inlet
    air cooling, International Journal of Energy Research 31:1292-1306.

3

Energy and Exergy Analysis of Reverse Brayton Refrigerator for Gas Turbine Power Boosting

Rahim K. Jassim[1], Majed M. Alhazmy[2] and Galal M. Zaki[2]
[1]*Department of Mechanical Engineering Technology,*
Yanbu Industrial College, Yanbu Industrial City
[2]*Department of Thermal Engineering and Desalination Technology,*
King Abdulaziz University, Jeddah
Saudi Arabia

1. Introduction

The output of Gas turbine (GT) power plants operating in the arid and semiarid zones is affected by weather conditions where the warm air at the compressor intake decreases the air density and hence reduces the net output power far below the ISO standard (15 °C and 60% relative humidity). The power degradation reaches an average of 7% for an increase in temperature by only 10°C above the 15 °C ISO standard. Furthermore; in hot summer days the plants are overloaded due to the increase in demand at peak periods, to meet the extensive use of air-conditioning and refrigeration equipment. The current techniques to cool the air at the compressor intake may be classified into two categories; direct methods employing evaporative cooling and indirect methods, where two loops refrigeration machines are used. Erickson (2003) reviewed the relative merits, advantages and disadvantages of the two approaches; Cortes and Willems (2003), and Darmadhkari and Andrepont (2004) examined the current inlet air cooling technology and its economic impact on the energy market.

In direct cooling methods water is sprayed at the compressor inlet bell mouth either through flexuous media (cellulose fiber) or fogging (droplets size in the order of 20 micron) into the air stream, Ameri *et al.* (2004). All spray cooling systems lower the intake temperature close to the ambient wet bulb temperature; therefore the use of the spray cooling is inefficient in coastal areas with high air humidity. Ameri *et al.* (2004) reported 13% power improvement for air relative humidity below 15% and dry bulb temperature between 31°C and 39°C. In addition to the effect of the ambient air humidity, the successful use of the direct method depends on the spray nozzles characteristics, Meher-Homji *et al* (2002) and droplets size, Bettocchi *et al.* (1995) and Meher-Homji and Mee, (1999). In evaporative cooling there is, to some extent, water droplets carry over problem, addressed by Tillman *et al* (2005), which is hazardous for compressor blades. Therefore, evaporative cooling methods are of limited use in humid coastal areas. Alhazmy *et al* (2006) studied two types of direct cooling methods: direct mechanical refrigeration and evaporative water spray cooler, for hot and humid weather. They calculated the performance improvement for ranges of ambient temperature and relative humidity, and their results indicated that the direct mechanical refrigeration increased the daily power output by 6.77% versus 2.5% for spray water cooling.

Indirect cooling by mechanical refrigeration methods can reduce the air temperature to any desirable value, even below the 15°C, regardless of the ambient humidity. There are two common approaches for air chilling: a) use of refrigeration units via chilled water coils, b) use of exhaust heat-powered absorption machines. As reported by Elliot (2001), application of mechanical air-cooling increases the net power on the expense of the thermal efficiency, 6% power boosting for 10°C drop in the inlet air temperature. Using absorption machines was examined for inlet air-cooling of cogeneration plants, Ondrays et al (1991), while Kakaras et al. (2004) presented a simulation model for NH_3 waste heat driven absorption machine air cooler. A drawback of the mechanical chilling is the risk of ice formation either as ice crystals in the air or as solidified layer on surfaces, such as the bell mouth or inlet guide vanes, Stewart and Patrick, (2000).

Several studies have compared the evaporative and mechanical cooling methods; with better performance for mechanical cooling. Mercer (2002) stated that evaporative cooling has increased the GT power by 10–15%, while the improvement for refrigeration chillers has reached 25%. Alhazmy and Najjar (2004) concluded that the power boosting varied between 1-7% for spray cooling but reached10-18% for indirect air cooling. In a recent study by Alhazmy et al (2006), they introduced two generic dimensionless terms (power gain ratio (PGR) and thermal efficiency change factor (TEC)) for assessment of intake air cooling systems. They presented the results in general dimensionless working charts covering a wide range of working conditions. Zadpoor and Golshan (2006) discussed the effect of using desiccant-based evaporative cooling on gas turbine power output. They have developed a computer program to simulate the GT cycle and the NOx emission and showed that the power output could be increased by 2.1%. In another development Erickson (2003 & 2005) suggested combination of the methods combining waste driven absorption cooling with water injection into the combustion air for power boosting; the concept was termed the "power fogger cycle". A novel approach has been presented by Zaki et al (2007), where a reverse Joule-Brayton air cycle was used to reduce the air temperature at the compressor inlet. Their coupled cycle showed a range of parameters, where both the power and thermal efficiency can be simultaneously improved.

As revealed in the above account abundant studies are concerned with the first law analysis of air intake cooling but those focuses on the second law analysis are limited. The basics of the second law analysis have been established and employed on variety of thermal systems by number of researchers. Bejan (1987, 1997), Bejan et al 1996, Rosen and Dincer (2003 a, and 2003 b) Ranasinghe, et al (1987), Zaragut et.al (1988), Kotas et al (1991) and Jassim (2003a, 2003b, 2004, 2005 and 2006), Khir et al (2007) have dealt extensively with various aspects of heat transfer processes. Chen et al (1997) analyzed the performance of a regenerative closed Brayton power cycle then extended the method to a Brayton refrigeration cycle, Chen et al (1999). The analysis considered all the irreversibilities associated with heat transfer processes. The exergy analysis of Brayton refrigeration cycle has been considered by Chen and Su (2005) to set a condition for the maximum exergetic efficiency while Tyagi et al (2006) presented parametric study where the internal and external irreversibilities were considered. The maximum ecological function, which was defined as the power output minus the power loss was determined for Brayton cycle by Huang et al (2000). Their exergy analysis was based on an ecological optimization criterion and was carried out for an irreversible Brayton cycle with external heat source.

The objective of the present analysis is to investigate the potential of boosting the power output of gas turbine plants operating in hot humid ambiance. The previously proposed coupled Brayton and reverse Brayton refrigeration cycles is analyzed employing both the energy and exergy analysis. Both the thermal and exergetic efficiencies are determined and the exergy destruction terms are evaluated.

2. Energy analysis

2.1 Brayton power cycle

Details of the first law energy analysis have been presented in a previous study by Zaki *et al* (2007) but the basic equations are briefly given here. Figure 1 shows the components of a coupled gas turbine cycle with Brayton refrigeration cycle. The power cycle is represented by states 1-2-3-4 and the reverse Brayton refrigeration cycle is represented by states 1-6-7-8-1. Portion of the compressed air $\alpha \dot{m}_1$ at pressure P_6 is extracted from the mainstream and cooled in a heat exchanger to T_7 then expands to the atmospheric pressure and T_8 as seen in Fig.1. The ambient intake air stream at T_o mixes with the cold stream at T_8 before entering the compressor.

Figure 2 shows the combined cycle on the T-s diagram; states o-$\overline{2}$-3-4 represent the power cycle without cooling, while the power cycle with air inlet cooling is presented by states 1-2-3-4-1. States 1-6-7-8 present the reverse Brayton refrigeration cycle.

In the mixing chamber ambient air at $\dot{m}_o$, T_o and ω_o enters the chamber and mixes with the cold air stream having mass flow rate of $\alpha \dot{m}_1$ at T_8. Air leaves the chamber at T_1, which depends on the ambient air conditions and the extraction ratio α.

The mass and energy balance for the mixing chamber gives the compressor inlet temperature as;

$$T_1 = \frac{(1-a)c_{po}T_o + a\,c_{p8}T_8}{c_{p1}} \tag{1}$$

Air leaves the compressor at $P_6 = xP_1$ flows through the Brayton refrigerator and the rest at $P_2 = rP_1$ as the working fluid for the power cycle, x is defined as the extraction pressure ratio and r is the pressure ratio.

The temperature of the air leaving the compressor at states 6 and 2 can be estimated assuming irreversible compression processes between states 1-2 and 1-6 and introducing isentropic compressor efficiency so that;

$$T_6 = T_1 + \frac{T_1}{\eta_{cx}}\left(x^{\frac{\gamma-1}{\gamma}} - 1\right) \tag{2}$$

and

$$T_2 = T_1 + \frac{T_1}{\eta_{cr}}\left(r^{\frac{\gamma-1}{\gamma}} - 1\right) \tag{3}$$

where, η_c is the compressor isentropic efficiency that can be estimated at any pressure z from Korakianitis and Wilson, (1994)

$$\eta_{cz} = 1 - \left(0.04 + \frac{z-1}{150} \right), \text{ where } z \text{ is either } x \text{ or } r \qquad (4)$$

The compression power between states 1 and 2 with extraction at state 6, separating the effects of the dry air and water vapor can be written as, Zaki *et al* (2007):

$$\dot{W}_c = \dot{m}_a \left[c_{pa} \frac{T_1}{\eta_{cr}} \left(r^{\frac{\gamma-1}{\gamma}} - 1 \right) + \omega_1 (h_{v2} - h_{v1}) + \left(\frac{a}{1-a} \right) \left(c_{pa} \frac{T_1}{\eta_{cx}} \left(x^{\frac{\gamma-1}{\gamma}} - 1 \right) + \omega_1 (h_{v6} - h_{v1}) \right) \right] \qquad (5)$$

where h_{vn} is the enthalpy of the saturated water vapor at the indicated state n. Equation 5 is a general expression for the compressor work for wet air; for dry air conditions $\omega_1 = 0$ and for stand alone GT $\alpha = 0$.

Heat balance about the combustion chamber gives the heat rate supplied by fuel combustion (net calorific value *NCV*) as:

$$\dot{Q}_{cc} = \dot{m}_f NCV = \left(\dot{m}_a + \dot{m}_f \right) c_{pg} T_3 - \dot{m}_a c_{pa} T_2 + \dot{m}_v (h_{v3} - h_{v2}) \qquad (6)$$

where h_{v2} and h_{v3} are the enthalpies of water vapor at the combustion chamber inlet and exit states respectively and f is the fuel to air ratio $f = \dot{m}_f / \dot{m}_a$ (related to the dry air flow rate). The total gases mass flow rate at the turbine inlet $\dot{m}_t = \dot{m}_a (1 + \omega_1 + f)$.

Substituting for T_2 from equation 3 gives the cycle heat input as:

$$\dot{Q}_{cc} = \dot{m}_a T_1 \left[(1+f) c_{pg} \frac{T_3}{T_1} - c_{pa} \left(\frac{r^{\frac{\gamma-1}{\gamma}} - 1}{\eta_{cr}} + 1 \right) + \frac{\omega_1}{T_1} (h_{v3} - h_{v2}) \right] \qquad (7)$$

The power produced by the turbine due to expansion of gases between states 3 and 4, Fig.1, is

$$\dot{W}_t = \dot{m}_a (1 + \omega_1 + f) c_{pg} \eta_t T_3 \left(1 - \frac{1}{r^{\frac{\gamma-1}{\gamma}}} \right) \qquad (8)$$

The turbine isentropic efficiency may be estimated using the practical relation recommended by Korakianitis and Wilson (1994) as:

$$\eta_t = 1 - \left(0.03 + \frac{r-1}{180} \right) \qquad (9)$$

Since, the gas turbine is almost constant volume machine at a specific rotating speed, the inlet air volumetric flow rate, $\dot{V}_a$ is fixed regardless of the intake air conditions. Equation 7 can be written in terms of the volumetric flow rate at the compressor inlet state by replacing

$\dot{m}_a$ by $\rho_a \dot{V}_a$. In the present analysis the moist air density ρ_a is assumed function of T_1 and ω_1. The Engineering Equation Solver (EES) software (Klein and Alvarado, 2004) has been used to calculate the wet air properties.

2.2 Brayton refrigeration cycle analysis

It is noted that the extraction pressure ratio x is the main parameter that determines the cold air temperature T_8 achievable by the air refrigeration cycle. The hot compressed air at P_6 and T_6 rejects its heat through a heat exchanger to cooling water. Similar to standard compressor intercoolers, state 7 will have the same pressure as P_6 for ideal case while T_8 depends on the cooling process. Irreversible expansion process between 7 and 8 with expansion efficiency η_{exp} yields,

$$T_8 = T_7 - T_7 \eta_{exp} \left[1 - \left(\frac{1}{x} \right)^{\frac{\gamma-1}{\gamma}} \right] \tag{10}$$

The power output of the expander, $\dot{W}_{exp}$ is

$$\dot{W}_{exp} = a \dot{m}_1 (h_7 - h_8) \tag{11}$$

The heat release $\dot{Q}_{out}$ between states 6 and 7 is computed by:

$$\dot{Q}_{out} = \alpha \dot{m}_1 (h_7 - h_6) \tag{12}$$

Since many of the desalination plants in the Gulf area use dual purpose combined GT units for water and power production, it is suggested here to utilize the rejected heat ($\dot{Q}_{out}$, Fig. 1) for brine heating or any industrial process that requires low grade heat. In general, the lower limit for T_7 is determined by the ambient temperature.

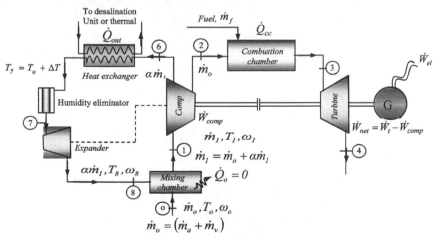

Fig. 1. A schematic of a gas turbine with an air cooler cycle

The cold stream flow rate $a\dot{m}_1$ proceeds to the mixing chamber to cool down the air at the compressor intake. The mixture temperature T_1 depends on the mass flow rate and the temperature of each stream as seen in Eq. 6.

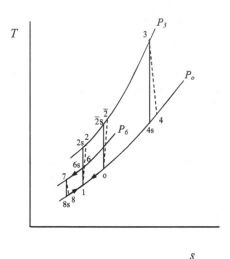

Fig. 2. T-s diagram for the proposed cycle

3. Exergy analysis

The exergy of a system is the maximum work obtainable as the system comes to equilibrium with the surrounding. The first law analysis (section 2) did not give information on heat availability at different temperatures. It is the second law which asserts that from engineering perspective. Therefore, exergy method is a technique based on both the first and second law of thermodynamics providing the locations where the exergy destruction and irreversibility are most. Ways and means can then be explored to reduce these exergy destructions to the practical minimum values.

The exergy destruction is a measure of thermodynamic imperfection of a process and is expressed in terms of lost work potential. In general, the expression for exergy destruction as articulated by Kotas (1995) is;

$$\dot{I} = T_o \left[\left(\dot{S}_{out} - \dot{S}_{in} \right) - \sum_{i=1}^{n} \frac{\dot{Q}_i}{T_i} \right] = T_o \dot{\Pi} \geq 0 \tag{13}$$

The exergy balance for an open system is,

$$\dot{E}_{in} + \dot{E}^Q = \dot{E}_{out} + \dot{W} + \dot{I} \tag{14}$$

where $\dot{E}^Q$ is the exergy associated with a given heat transfer rate, is termed as thermal exergy flow (kW). Quantities of lost work or rate of exergy destruction due to irreversibility can be estimated for each component of the coupled cycles as follows.

3.1 Mixing chamber

The rate of exergy destruction in the mixing chamber includes different forms of exergy; physical due to changes in the properties during mixing and that due to chemical reactions if any. The *physical exergy* of a mixture of N components can be evaluated from Jassim, *et al* 2004 as,

$$\left(\tilde{\varepsilon}_{ph}\right)_M = \sum_{j=1}^{N} y_j \; \tilde{\varepsilon}_j^{\Delta T} + \tilde{R}T_o \; \ell n\left(\frac{P}{P_o}\right) \tag{15a}$$

where P is the total pressure of the mixture. The equation may be written in another form, using tabulated values of properties ($\tilde{c}_p^h$ and $\tilde{c}_p^s$) as;

$$\left(\tilde{\varepsilon}_{ph}\right)_M = \sum_{j=1}^{N} y_j \left[\tilde{c}_p^h \left(T - T_o\right) - T_o \; \tilde{c}_p^s \; \ell n\left(\frac{T}{T_o}\right) \right]_j + \tilde{R}T_o \; \ell n\left(\frac{P}{P_o}\right) \quad \text{(kJ/kmol)} \tag{15b}$$

The mean physical exergies for the mixing streams are calculated using equation 15b. The values of mean molar isobaric exergy capacities for enthalpy $\left(\tilde{c}_{p,j}^h\right)$ and entropy $\left(\tilde{c}_{p,j}^s\right)$ have been evaluated from (Kotas, 1995, Table D3).

The chemical exergy of a mixture $\tilde{\varepsilon}_{oM}$ due to changes in number of moles is given by the following expression (Kotas, 1995)

$$\tilde{c}_{oM} = \sum_j y_j \tilde{c}_{oj} + \tilde{R}T_o \sum_j y_j \; \ell n \; x_j \quad \text{(kJ/kmol)} \tag{16}$$

where y_j is the mole fraction of component j in a mixture

$\tilde{\varepsilon}_o$ is the molar chemical exergy (kJ/kmol) evaluated from (Kotas, 1995, Table D3).

The irreversibility involved in mixing hot and cold air streams as shown in Fig. 1 assuming adiabatic mixing process ($\dot{Q}_o = 0$) is,

$$\dot{I}_{mc} = \dot{E}_8 + \dot{E}_o - \dot{E}_1 \tag{17a}$$

or

$$\dot{I}_{mc} = a\dot{m}_1 \, \varepsilon_8 + \dot{m}_o \, \varepsilon_o - \dot{m}_1 \, \varepsilon_1 \tag{17b}$$

In general, the specific exergy component for each of the terms in Eq. 17b is,

$$\varepsilon = \varepsilon_o + \varepsilon_{ph} \tag{17c}$$

The values of ε_{ph} and ε_o can be calculated from Eqs. (15b) and (16), respectively. Hence, equations 17b and 17c give the total rate of exergy losses in the mixing chamber as,

$$\dot{I}_{mc} = a\dot{m}_1 \left(\varepsilon_{o,8} + \varepsilon_{ph,8}\right) \Big/ \tilde{m}_8 + \dot{m}_o \left(\varepsilon_{o,o} + \varepsilon_{ph,o}\right) \Big/ \tilde{m}_o - \dot{m}_1 \left(\varepsilon_{o,1} + \varepsilon_{ph,1}\right) \Big/ \tilde{m}_1 \tag{17d}$$

$$\dot{I}_{mc} = \dot{m}_a \left(1 + \omega_1\right) \left[\left(\frac{a}{1-a}\right) \frac{\left(\varepsilon_{o,8} + \varepsilon_{ph,8}\right)}{\tilde{m}_8} + \frac{\left(\varepsilon_{o,o} + \varepsilon_{ph,o}\right)}{\tilde{m}_o} - \left(\frac{1}{1-a}\right) \frac{\left(\varepsilon_{o,1} + \varepsilon_{ph,1}\right)}{\tilde{m}_1} \right]$$

3.2 Compressor

The rate of exergy destruction in the compressor (see Figs 1 and 2) for the compression process 1-2, is,

$$\dot{I}_c = \dot{m}_o \, T_o \left(s_2 - s_1\right) + a \left(\frac{\dot{m}_o}{1-a}\right) T_o \left(s_6 - s_1\right) \tag{18}$$

Substituting the entropy change for gases, Cengel and Bolos, 2005 gives

$$\dot{I}_c = \dot{m}_a \left(1 + \omega_1\right) T_o \left[c_{pa} \, \ell n \left(\frac{T_2}{T_1}\right) - R_a \, \ell n \left(\frac{P_2}{P_1}\right) + \left(\frac{a}{1-a}\right) \left(c_{pa} \, \ell n \left(\frac{T_6}{T_1}\right) - R_a \, \ell n \left(\frac{P_6}{P_1}\right) \right) \right] \tag{19}$$

The mixing chamber irreversibility is added to the compressor irreversibility as it is caused by air mixing before entering the compressor. Therefore the compressor effective power is,

$$\dot{W}_{eff,c} = \dot{W}_c + \dot{I}_c + \dot{I}_{mc} \tag{20}$$

3.3 Combustion chamber

The combustion process irreversibility is often accomplished by heat transfer, fluid friction, mixing and chemical reaction. In principle it is difficult to evaluate an absolute value for the contribution of each process to the total irreversibility. However, the process of combustion can be examined by assuming that it takes place under adiabatic conditions and neglect irreversibilities due to friction and mixing. The important term of fuel combustion is the chemical exergy ε_o which has been extensively dealt with by Kotas, 1995. The exergy due to chemical reaction of a combustible component (k) is function of other parameters as;

$$\varepsilon_o = f(\Delta h, \Delta s, T_o, X_k) \tag{21}$$

The subscript k refers to the component's mass (mole) fraction of products composition.

The chemical exergy ε_o may be expressed as percentage of the net calorific value (NCV) as,

$$\varepsilon_o = \varphi \, NCV \tag{22}$$

The value of φ is function of the fuel components mass to Carbon ratio, (O/C) for Oxygen, (H/C) for hydrogen and (N/C) for Nitrogen. For liquid fuels additional ratio for the sulfur content is included (S/C), and φ is in the order of 1.04 to 1.08. Typical values of φ for some industrial fuels are given in Kotas 1995. Therefore, the rate of exergy loss $T_o \Delta S_o$ in a combustion reaction is;

$$T_o \Delta S_o = \dot{m}_f \times NCV \left(\varphi - 1\right) = \dot{m}_a \times f \times NCV \left(\varphi - 1\right) \tag{23}$$

The exergy destruction rate during the combustion process is the entropy change between states 2 and 3 due to chemical combustion reaction is;

$$\dot{I}_{cc} = T_o\left[(S_P)_3 - (S_R)_2\right] \tag{24a}$$

where $(S_R)_2 = (S_a)_2 + (S_f)_2$ and the subscripts P, R, a, and f represent *products, reactants, air* and *fuel*, respectively (Nag, 2002). Then

$$\dot{I}_{cc} = T_o\left[(S_P)_3 - \left[(S_a)_2 + (S_f)_2\right]\right] \tag{24b}$$

Introduce the entropy difference at the reference state ΔS_o as

$$\Delta S_o = (S_P)_o - \left[(S_f)_o + (S_a)_o\right] \tag{24c}$$

The exergy destruction rate becomes

$$\dot{I}_{cc} = T_o\left\{\left[(S_P)_3 - (S_P)_o\right] - \left[(S_a)_2 - (S_a)_o\right] + \Delta S_o\right\} \tag{24d}$$

Substitute for the entropy change for reactants and products gives;

$$\dot{I}_{cc} = \dot{m}_a T_o \left\{ \begin{array}{l} (1+f+\omega_1)\left[c_{pg}\,\ell n\left(\dfrac{T_3}{T_o}\right) - R_g\,\ell n\left(\dfrac{P_3}{P_o}\right)\right] - \\[4mm] (1+\omega_1)\left[c_{pa}\,\ell n\left(\dfrac{T_2}{T_o}\right) - R_a\,\ell n\left(\dfrac{P_2}{P_o}\right)\right] \end{array} \right\} + T_o \Delta S_o \tag{24e}$$

where $R_g = \dfrac{c_{pg}(\gamma-1)}{\gamma}$

The total energy input $\dot{Q}_{eff,cc}$ to the combustion chamber is then

$$\dot{Q}_{eff,cc} = \dot{Q}_{cc} + \dot{I}_{cc} \tag{25}$$

3.4 Gas turbine

The rate of exergy destruction in the gas turbine is,

$$\dot{I}_t = \dot{m}_a\left(1+f+\omega_1\right)T_o\left(s_4 - s_3\right)$$

$$= \dot{m}_a\left(1+f+\omega_1\right)T_o\left[c_{pg}\,\ell n\left(\dfrac{T_4}{T_3}\right) - R_g\,\ell n\left(\dfrac{P_4}{P_3}\right)\right] \tag{26}$$

The effective power output of the turbine can be expressed in terms of exergy destruction rate as;

$$\dot{W}_{eff,t} = \dot{W}_t - \dot{I}_t \tag{27}$$

3.5 Heat exchanger

During the air cooling process 6-7 (Fig. 1), the rate of exergy destruction in the heat exchanger is,

$$\dot{I}_{HX} = a\dot{m}_1 T_o \left(s_6 - s_7\right) + \frac{T_o}{T_6}\dot{Q}_{out} \tag{28}$$

where $\dot{Q}_{out}$ is the rate of heat rejection from the heat exchanger at T_6, $= a\dot{m}_1 \left(h_6 - h_7\right)$.

The net energy output in the HX is then;

$$\dot{Q}_{eff,HX} = \dot{Q}_{out} + \dot{I}_{HX} \tag{29}$$

3.6 Expander

The rate of exergy destruction in the expander (process 7-8) is similar to that in the turbine and can be expressed as,

$$\dot{I}_{exp} = a\dot{m}_1 T_o \left(s_8 - s_7\right)$$

$$= a\left(\frac{\dot{m}_a}{1-a}\right)(1+\omega_1)T_o \left[c_{pa} \, \ell n \left(\frac{T_8}{T_7}\right) - R_a \, \ell n \left(\frac{P_8}{P_7}\right) \right] \tag{30}$$

The effective power output of the expander is then

$$\dot{W}_{eff,exp} = \dot{W}_{exp} - \dot{I}_{exp} \tag{31}$$

To that point the exergy destruction have been evaluated for all the system components and the total rate of exergy losses in the plant, Fig. 1 is;

$$\sum \dot{I} = +\dot{I}_{mc} + \dot{I}_c + \dot{I}_{cc} + \dot{I}_t + \dot{I}_{HX} + \dot{I}_{exp} \tag{32}$$

4. Performance of the integrated system

For the proposed cycle the net power output and heat input, can be easily calculated using equations 5, 8 and 11. If the expander power $\dot{W}_{exp}$ is recovered, then the net output of the cycle may be expressed as:

$$\dot{W}_{net,with cooling} = \dot{W}_t - \left(\dot{W}_{comp} - \dot{W}_{exp}\right) \tag{33}$$

An advantage of the present cycle is the availability of useful energy $\dot{Q}_{out}$ that may be utilized for any process. Let us define a term for the useful heat input as the net heat utilized to produce the shaft power, therefore, equations 7 and 12 give,

$$\dot{Q}_{useful} = \dot{Q}_{cc} - \dot{Q}_{out} \tag{34}$$

An integral useful thermal efficiency $\left(\eta_{th,u}\right)$ may be defined as:

$$\eta_{th,u} = \frac{\dot{W}_{net}}{\dot{Q}_{useful}} \tag{35}$$

If the GT is in operation without the cooling cycle then $\dot{Q}_{out} \to 0$ and $\dot{W}_{exp} \to 0$, equation 35 leads to the conventional thermal cycle efficiency expression (without cooling $\eta_{th,nocooling}$).

The net power for the GT unit without cooling is obtained by introducing $\alpha = 0$ and $\omega_1 = \omega_o$ in Equation 5 to get $\dot{W}_{c,no\ cooling}$. Similarly the turbine power without cooling $\dot{W}_{t,no\ cooling}$ is obtained by subsisting $\omega_1 = \omega_o$, $T_1 = T_o$ and f is calculated from Eq. 8 using $T_{\bar{2}}$ instead of T_2 Therefore,

$$\dot{W}_{net,no\ cooling} = \left| \dot{W}_t - \dot{W}_c \right|_{no\ cooling} \tag{36}$$

Adopting the terminology proposed by Alhazmy *et al* (2006) and employed in our previous work Zaki, *et al* (2007) to evaluate the thermal efficiency augmentation; a thermal efficiency change term based on the useful input energy TEC_u is defined as:

$$TEC_u = \frac{\eta_{th,u} - \eta_{th,no\ cooling}}{\eta_{th,no\ cooling}} \times 100\% \tag{37}$$

The subscript u refers to useful indicating that the GT plant is serving adjacent industrial process so that the conventional thermal efficiency definition is not applicable for this condition. For the present parametric analysis let us focus on a simple GT plant just coupled to a Brayton refrigerator without use of the cooling energy; the term $\dot{Q}_{out}$ in Eqs. 34 is eliminated and the performance of the coupled cycles may be determined from a power gain ratio (PGR) and thermal efficiency change (TEC) terms as:

$$PGR = \frac{\dot{W}_{net,with\ cooling} - \dot{W}_{net,no\ cooling}}{\dot{W}_{net,no\ cooling}} \times 100\% \tag{38}$$

$$TEC = \frac{\eta_{th,with\ cooling} - \eta_{th,no\ cooling}}{\eta_{th,no\ cooling}} \times 100\% \tag{39}$$

The PGR is a generic term that takes into account all the parameters of the gas turbine and the coupled cooling system. For a stand-alone gas turbine under specific climatic conditions; PGR = 0. If Brayton refrigerator is used, the PGR increases with the reduction of the intake temperature. However, the PGR gives the percentage enhancement in power generation; the TEC of a coupled system is an important parameter to describe the fuel utilization effectiveness.

From the second law analysis the performance of a system may be articulated by the exergetic efficiency. For the proposed system define the exergetic efficiency η_{ex} as a ratio of total exergy output $\left(\dot{E}_{out}\right)$ to input $\left(\dot{E}_{in}\right)$ rates as;

$$\eta_{ex} = \frac{\dot{E}_{out}}{\dot{E}_{in}} \qquad (40)$$

Exergy balance for the system shown in Figure (1) can be expressed in the following form,

$$\dot{E}_{out} = \dot{W}_{eff,t} + \dot{W}_{eff,exp} - \dot{W}_{eff,c} \qquad (41)$$

The total rate of exergy input is

$$\dot{E}_{in} = \dot{Q}_{eff,cc} - \dot{Q}_{eff,HX} \qquad (42)$$

Expressions for the effective power terms $\dot{W}_{eff,c}$, $\dot{W}_{eff,t}$, and $\dot{W}_{eff,exp}$ are given by equations 20, 27 and 31 respectively, while $\dot{Q}_{eff,cc}$ and $\dot{Q}_{eff,HX}$ are determined from equations 25 and 29.

The irreversibilities of heat exchangers $\left(\dot{I}_{HX}\right)$, expanders $\left(\dot{I}_{exp}\right)$ and mixing chamber $\left(\dot{I}_{mc}\right)$ are all equal to zero when the GT operates on a stand alone mode.

Similar to the criteria adopted for the energy analysis let us define dimensionless terms as the *exergetic power gain ratio* (PGR$_{ex}$) and *exergetic thermal efficiency change* (TEC$_{ex}$) as:

$$PGR_{ex} = \frac{\left(\dot{E}_{out}\right)_{cooling} - \left(\dot{E}_{out}\right)_{nocooling}}{\left(\dot{E}_{out}\right)_{nocooling}} \times 100\% \qquad (43)$$

$$TEC_{ex} = \frac{\eta_{ex,cooling} - \eta_{ex,nocooling}}{\eta_{ex,nocooling}} \times 100\% \qquad (44)$$

5. Results and discussion

In order to evaluate the integration of Brayton refrigerator for intake air-cooling on the GT performance, a computer program has been developed using EES (Klein and Alvarado 2004) software. The computation procedure was first verified for the benchmark case of simple open cycle with dry air as the working fluid, for which $\alpha = 0$, $\omega_o = 0$ $f = 0$ and assuming isentropic compression and expansion processes, Eqs. 5, 7, 8 and 37 leads to the expression for the thermal efficiency of the air standard Brayton cycle.

$$\eta_{th} = 1 - \frac{1}{r^{(\gamma-1)/\gamma}} \qquad (45)$$

Similarly from the exergy point of view and for the same ideal conditions, all exergy destruction terms (Equations 17d, 24e, 26, 28, 30) and the exergetic efficiency (equation 44) leads to the same expression, Eq. 45.

For the present analysis, the ambient air at Jeddah, Saudi Arabia (Latitude 22.30° N and Longitude 39.15° E) a typical city with over 40 GT plants operating under sever weather conditions was selected (ambient temperature of T_o = 45°C and φ_o = 43.3%). The air temperature at the expander inlet T_7 is determined by the requirements of the heating process; the coolant flow rate can be controlled to obtain different outlet temperature suitable for the industrial process. For 10K terminal temperature difference the value of T_7 can be fixed to 55°C. Table I shows the range of parameters for the present analysis.

In the previous study by the authors Zaki et al (2007), the effect of the extraction pressure ratio (x) and the ratio between the mass rates (α) has been investigated. It has been established that for a fixed extraction pressure (P_6), increasing (α) increases the chilling effect and hence T_1 decreases as seen in Fig. 4. It is possible to adjust the values of α and x to operate the GT close to the ISO standard (288.15 K the dotted line in Figure 4). It is seen that a reasonable range of operation is bounded by $2 < x < 4$ and $0.2 < \alpha < 0.3$ for which the intake temperature drops from 318 K to values between 310 and 288 K. Extending the range to reach temperatures below 273 K is constrained by frost formation at the compressor mouth and expander outlet.

Parameter	Range
Ambient air	
Max. ambient air temperature, T_a	318.15 K
Relative humidity, RH_o	43.4 %
Volumetric air flow rate, $\dot{V}_a$	1 m^3 s^{-1}
Net Calorific Value, NCV	42500 kJ kg^{-1}
Reference ambient temperature, T_o	298.15 K
Gas Turbine	
Pressure ratio P_2/P_1	10
Turbine inlet temperature T_3	1373.15 K
Specific heat ratio of gas, γ	1.333
Specific heat of gas, c_{pg}	1.147 kJ kg^{-1} K^{-1}
Air Compressor	
Standard environment temp., T_o	298.15 K
Extraction pressure ratio P_x/P_1	2 → 9
Extracted mass ratio, a	0.1 → 0.5
Specific heat ratio of air, γ	1.4
Heat exchanger	
ΔT	10 K
Fuel	
Fuel oils, ϕ	1.08

Table 1. Range of parameters for the present analysis

Figure 5 shows the *PGR* and *TEC* variations within pressure extraction ratio up to 4 and $0 \leq \alpha \leq 0.5$. It is seen that the power gain increases linearly with α, which means enhanced chilling effect due to the large air mass rate ($\dot{m}_a \alpha$) passing through the refrigeration cycle. As seen in the figure the power is boosted up while the thermal efficiency (Eq. 39) decreases, at $x = 3$ and 4, due to the increase in fuel consumption rate. The drop in the *TEC* is quite large for high α and x values. For moderate values as $\alpha = 0.2$ and extraction pressure of 4 bar, the power increases by 9.11%, while the thermal efficiency drops by only 1.34%. This result indicates that passing 20% of the intake air at 4 bars pressure thorough a Reverse Brayton refrigerator increases the power by 9.11% with only drop in thermal efficiency of 1.34%. Therefore, selection of the operation parameters is a trade off between power and efficiency, boosting the power on penalty of the thermal efficiency. It is of interest to determine the parameters at which the power is boosted while the thermal efficiency of the GT stays the same as that for the condition of a stand alone operation. This means that the *TEC* (Eq. 39) approaches zero. Solving Eq. 39 shows that within a narrow range of extraction pressure ($x = 2.6$ to 2.8) as shown in Figure 3, the plant thermal efficiency is slightly altered from the stand alone efficiency and *TEC* is independent on the extracted mass ratio α. Though, the efficiency does not change appreciably the power gain strongly depends on α.

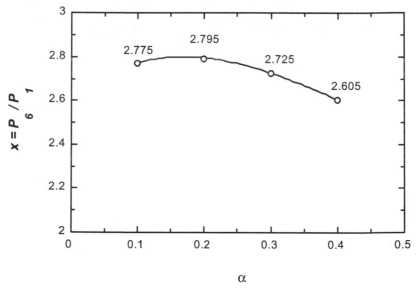

Fig. 3. Solution of Eq. 39, *TEC* = 0 gives the conditions for constant thermal efficiency

The change of the *PGR* within a range of pressure ratios is presented in Fig 6 a, indicating increase in *PGR* up to certain α followed by a descending trend. For a constant value of α the power gain increases with the extraction pressure reaching a peak value then decreases. As seen in Fig. 6 b. the increase in x or P_6 while T_7 is constant means that the outlet temperature from the expander moves from T_8 to $T_{\bar{8}}$, which means lower intake temperature ($T_{\bar{8}} < T_8$) with tendency to increase the *PGR*. On the other hand increase in x results high net compressor work ($\dot{W}_{comp} - \dot{W}_{exp}$) in Equation 33. High net compressor work reduces the nominator of Eq. 38 and decreases the *PGR*. Therefore, the *PGR* ascending descending trend in Figure 6 a. is expected.

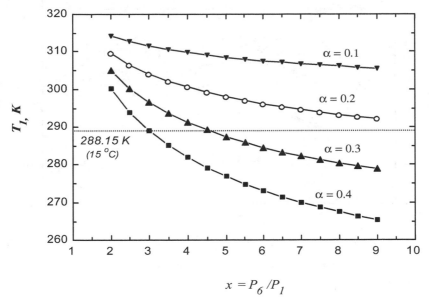

$$x = P_6 / P_1$$

Fig. 4. Effect of extraction pressure x and mass ratio α on the gas turbine intake air temperature.

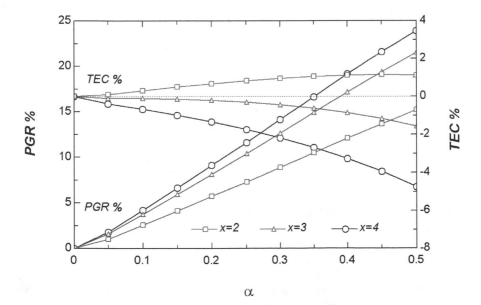

Fig. 5. Power enhancement (Eq.38) and thermal efficiency change (Eq. 39) for GT plant, $r = 10$

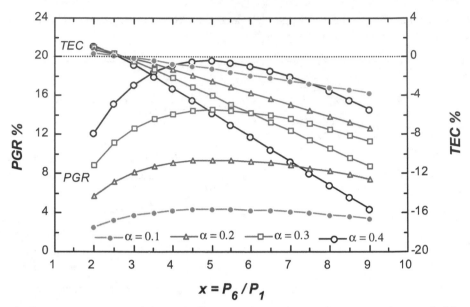

Fig. 6a. Power gain ratio and thermal efficiency change factors for a gas turbine cooled by air Brayton refrigerator

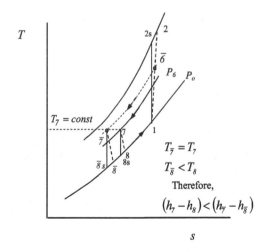

Fig. 6b. *T-s* diagram to illustrate the effects of increasing the extraction pressure. Increasing the extraction pressure P_6 reduces the compressor intake temperature and increases the difference ($\dot{W}_{comp} - \dot{W}_{exp}$).

To illustrate the effect of the heat rate ($\dot{Q}_{out}$) during the cooling process (process 6-7, Fig. 1) on the overall system performance, the useful thermal efficiency change (TEC_u, Eq. 37) is plotted with the control variables α and x in Figure 7. It is seen that for all operation conditions TEC_u is positive, which means that the energy utilized for an industrial process has been deducted from the heat released by fuel. Though the result was expected, the values here can not be compared to the stand alone plant or even with the case where the cooling water energy is wasted. The figure shows also the *PGR* variation, which is the same whether the energy is utilized or wasted as indicated by Equations 35 and 38.

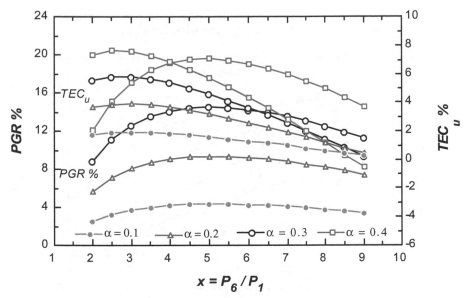

Fig. 7. The useful thermal efficiency change as function of α and x

The results of the energy analysis showed clearly the advantages of the proposed cooling system and the expected magnitude of performance enhancement. However, the practical range of operation parameters was determined the contribution of each of the system components to the irreversibility is only determined through equations 17d, 19, 24e, 26, 28 and 30. Table 2 shows numerical values for the exergy rate of the different components at $x = 3.5$ and variable α.

a	$\dot{I}_c$	$\dot{I}_{cc}$	$\dot{I}_t$	$\dot{I}_{HX}$	$\dot{I}_{exp}$	$\dot{I}_{mc}$	$\dot{I}_{total}$
0.00	21.4	360.8	22.7	0	0	0	0
0.05	21.76	369.9	22.97	17.39	0.3103	0.6484	433
0.10	22.18	381.3	23.3	35.68	0.6644	1.377	464.5
0.15	22.65	393.2	23.64	54.95	1.07	2.133	497.6
0.20	23.2	405.4	24.00	75.28	1.539	2.917	532.4
0.25	23.83	418.2	24.36	96.77	2.082	3.733	568.9
0.30	24.56	431.4	24.74	119.5	2.718	4.581	607.5
0.35	25.42	445.1	25.12	143.8	3.468	5.464	648.3
0.40	26.44	459.3	25.52	169.6	4.361	6.386	691.6
0.45	27.66	474.1	25.94	197.3	5.437	7.349	737.8
0.50	29.14	489.5	26.36	227.1	6.753	8.356	787.2

Table 2. Cycle components irreversibility values (kW) for x =3.5

It is seen that the major contribution comes from the combustion chamber where the irreversibility is the highest and presents 62 to 85% of the total irreversibility. The heat exchanger comes next with nearly 5 % to 46% of the magnitude of the combustion chamber. Both the compressor and turbine are of nearly equal magnitude. Therefore, optimization of the system should focus on the irreversibilities in the combustion chamber and heat exchanger. The mixing chamber irreversibility is found to be very small compared to the other components as seen in Table 2 and can be ignored in future studies.

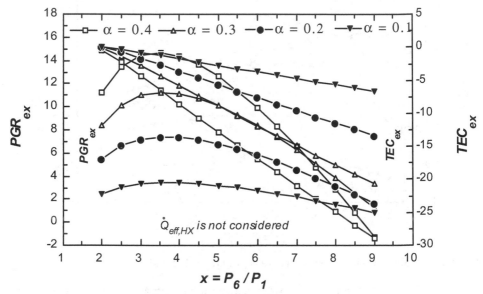

Fig. 8. Dependence of exergetic power gain ratio and exergetic thermal efficiency on the extracted pressure ratio without taking $\dot{Q}_{eff,HX}$ into account

The variations of the exergetic PGR and exergetic TEC (equations 43 and 44) are shown in Figure 8 for the same range of parameters as in Fig. 6 a. Clearly this figure shows that the irreversibilities give lower values for the power gain and efficiency as compared to the energy analysis. The difference, at the maximum PGR, conditions is in the order of 20 to 25% as seen in Table 3. For example, the maximum PGR is reduced from 9.37 in case of first law of analysis to 7.436 when the second law was considered.

α	$(PGR)_{max}\%$ Fig. 6 a	$(PGR_{ex})_{max}\%$ Fig. 8	Reduction %
0.1	4.336	3.448	20.478
0.2	9.375	7.436	20.683
0.3	14.5	11.25	22.414
0.4	19.58	14.66	25.128

Table 3. Effect of irreversibility on power gain ratio

The energy destruction terms in Table II show dependency on α at constant pressure ratio x. The difference between the gain in power as predicted by the energy and exergy analysis is affected by both the amount of air that circulate in the refrigeration loop and the coupling pressure ratio x. Figure 9 indicates clearly that the difference increases at high extraction ratio (more air is bleed from the main compressor) and high extraction pressure (large portion of the energy is wasted in the heat exchanger). Within the previously selected range of parameters $2 < x < 4$ and $0.2 < \alpha < 0.3$ the maximum difference is in the order of 5%.

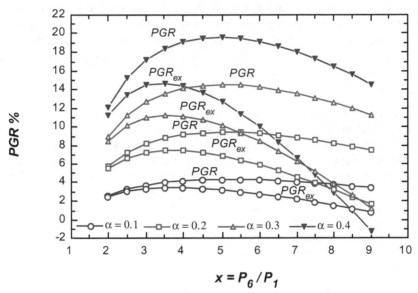

Fig. 9. Energetic and exergetic power gain ratio for a gas turbine cooled by air Brayton refrigerator

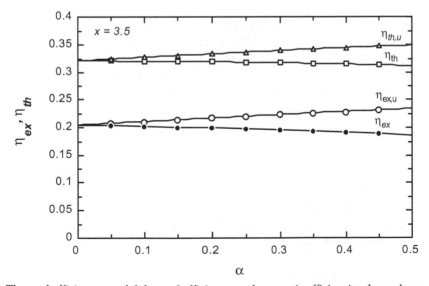

Fig. 10. Thermal efficiency, useful thermal efficiency and exergetic efficiencies dependency on α.

Figure 10 shows the magnitudes of the thermal efficiency, the useful thermal efficiency (equation 35) and the exergetic thermal efficiency (equation 40) for $0 \leq \alpha \leq 0.5$ and extraction pressure of 3.5 bars. It is clear that utilization of the cooling energy for any process enhances the use of the input fuel for which $\eta_{th,u} > \eta_{th}$ and $\eta_{ex,u} > \eta_{ex}$ as seen in the figure. In general, the thermal efficiency weights all thermal energy equally, whilst the exergetic efficiency acknowledges the usefulness of irreversibilities on its quality and quantity. Thus, exergetic efficiency is more suitable for determining the precise power gain ratio. The figure shows also that the useful efficiency whether energetic or exergetic is higher than the corresponding values when wasting the heat exchanger heat rejection $\dot{Q}_{out}$.

6. Conclusions

In this study a new approach for boosting the power of gas turbine power plants by cooling the intake air is analyzed by the energy and exergy methods. The gas turbine inlet temperature is reduced by mixing chilled air from a Brayton refrigeration cycle and the main intake air stream. The air intake temperature depends on two parameters, the cold air stream temperature from the reverse Brayton cycle and the ambient hot humid air conditions. The energy analysis of the coupled Brayton-reverse Brayton cycles showed that the intake air temperature could be reduced to the ISO standard (15°C) and the gas turbine performance can be improved. This study demonstrated the usefulness of employing exergy analysis and the performance improvement was expressed in terms of generic dimensionless terms, *exergetic power gain ratio (PGRex)* and *exergetic thermal efficiency change (TECex)* factor.

The performance improvement of a GT irreversible cycle of 10 pressure ratio operating in hot weather of 45°C and 43.4% relative humidity was investigated for extraction pressures from 2 to 9 bars and cold to hot air mass rate ratio from 0.1 to 0.5. The results showed that the combustion chamber and the cooling heat exchanger are the main contributors to the exergy destruction terms; the combustion chamber irreversibility was the highest and presented 62 to 85% of the total irreversibility. The heat exchanger comes next with nearly 5 % to 46% of the combustion chamber. The irreversibility of mixing chamber was found to be small compared to other components and can be safely ignored. On basis of the energy analysis the GT power can be boosted up by 19.58 % of the site power, while the exergy analysis limits this value to only 14.66% due to exergy destruction in the components of the plant. The irreversibility can be reduced by optimal design of the combustion chamber, the heat exchanger and selecting optimum operational parameters of the coupled power and refrigeration units.

7. Nomenclature

c_{pa}	= specific heat of air at constant pressure, $kJ\ kg^{-1}\ K^{-1}$
c_{pg}	= specific heat of gas at constant pressure, $kJ\ kg^{-1}\ K^{-1}$
$\dot{E}$	= total exergy rate, kW
f	= fuel to air ratio
$\dot{i}$	= rate of exergy destruction, kW
h	= specific enthalpy, $kJ.kg^{-1}$

$\dot{m}$	= mass flow rate, $kg\ s^{-1}$	
NCV	= net calorific value, $kJ\ kg^{-1}$	
P	= pressure, kPa	
PGR	= power gain ratio	
PGR_{ex}	= exergetic power gain ratio	
$\dot{Q}$	= heat rate, kW	
$\dot{Q}_{out}$	= heat release rate from the condenser, kW	
r	= pressure ratio = P_2/P_1	
s	= specific entropy, $kJ\ kg^{-1}K^{-1}$	
T	= temperature, K	
TEC	= thermal efficiency change factor	
TEC_u	= useful thermal efficiency change factor	
TEC_{ex}	= exergetic thermal efficiency change	
$\dot{W}$	= output power (according to subscript), kW	
x	= extraction pressure ratio, P_6/P_1	
y	= mole fraction of fuel components	

Greek symbols

α	= fraction of air mass flowing through the cooler cycle
ε	= specific exergy (according to subscript)
ε_o	= exergy due to chemical reaction, $kJ\ mol^{-1}$
ω	= specific humidity, kg water vapor/kg dry air
ω_o	= ambiant specific humidity, kg water vapor/kg dry air
η	= efficiency (according to subscript)
γ	= specific heat ratio

Subscripts

o	= ambiant
a	= dry air
c	= compressor
cc	= combustion chamber
cy	= cycle
ex	= exergy
exp	= expander
g	= gas
HX	= heat exchanger
p	= products
pg	= product gas
mc	= mixing chamber
R	= reactants
t	= turbine
u	= useful
v	= water vapor

Superscripts

$\sim$	= mean molar value of a property

8. References

Alhazmy MM, Jassim RK, Zaki GM. 2006. Performance enhancement of gas turbines by inlet air-cooling in hot and humid climates. *International Journal of Energy Research* 30:777-797

Alhazmy MM, Najjar YSH. 2004. Augmentation of gas turbine performance using air coolers. *Applied Thermal Engineering* 24:415-429.

Ameri M, Nabati H, Keshtgar A. 2004. Gas turbine power augmentation using fog inlet cooling system. *Proceedings ESDA04 7th Biennial Conference Engineering Systems Design and Analysis*, Manchester UK., paper ESDA2004-58101.

Bejan, A, 1987, "The thermodynamic design of heat transfer: Second Law Analysis of Thermal Systems, proceedings of the Fourth Int. symposium on Second Law Analysis of Thermal Systems, Rome, Italy,1-15 ASME, N.Y.

Bejan, A, Tsatsaronis, G., and Moran, M., 1996, "*Thermal design and optimization*" John Wiley, New York, USA.

Bejan, A,1997, "*Advanced engineering thermodynamics*" John Willey, New York, USA.

Bettocchi R, Spina PR, Moberti F. 1995. Gas turbine inlet air-cooling using non-adiabatic saturation, *Process, ASME Cogen-Turbo power Conference*, paper 95-CTP-49:1-10.

Cengel Y A , Bolos M A, 2005. Thermodynamics: An engineering approach (6th edition). McGraw Hill New York.

Chen, C.K. Su, Y.F. (2005) 'Exergetic efficiency optimization for an irreversible Brayton refrigeration cycle'. Int. J. of thermal sciences, 44 (3) pp 303-310

Chen, L, Ni, N, Wu, C, Sun, F, 1999, Performance analysis of a closed regenerated Brayton heat pump with internal irreversibilities, Journal of Energy Res. 23 pp 1039-1050.

Chen, L, Sun, F, Wu, C. Kiang, R.L.1997, Theoretical analysis of the performance of a regenerative closed Brayton cycle with internal irreversibilities. Energy Convers. Mangt. 38, pp 871-877.

Cortes C.P.E. and Willems D. (2003) 'Gas turbine inlet cooling techniques: An overview of current technology', *Proceedings Power GEN International 2003*, Las Vegas Nevada, USA, 9-11 December, paper 13B-4, pp 314-321.

Darmadhikari S. and Andrepon J.S. (2004). 'Boost gas turbine performance by inlet air cooling', *J. Hydrocarbon Processing*, 83 (2, pp 77-86.

Elliot J. (2001) 'Chilled air takes weather out of equation', *Diesel and gas turbine worldwide*, V 33 # 9 October/November 2001, pp 94-96.

Erickson DC, 2003. Aqua absorption turbine inlet cooling, *Proceedings of IMEC 03, ASME International Mechanical Engineering Congress & Exposition*, Washington DC 16-21, November.

Erickson DC, 2005. Power fogger cycle, *ASHRAE Transactions*, 111, part 2:551-554.

Huang, Y.C., Hung, C.I.,Chen, CK, 2000 "An ecological exergy analysis for an irreversible combined Brayton engine with an external heat source. Journal of Power and Energy, 214, No 1, pp 413-421.

Jassim RK, 2003b, "Evaluation of combined heat and mass transfer effect on the thermoeconomic optimisation of an air-conditioning rotary regenerator" Trans, ASME, Journal of Heat Transfer, Vol.125, No.4, pp 724-733.

Jassim, RK, Habeebullah, B A, Habeebullah, AS, (2004) "Exergy Analysis of Carryover Irreversibilities of a Power Plant Regenerative Air Heater". Journal of Power and Energy, 218, No 1, pp 23-32.

Jassim, RK, Khir, T, Ghaffour, N, 2006, "Thermoeconomic optimization of the geometry of an air conditioning precooling air reheater dehumidifier", Int. Journal of Energy Research, Vol. 30, pp 237-258.

Jassim, RK, Khir, T, Habeebullah, BA, Zaki, GM, 2005, "Exergoeconomic Optimization of the Geometry of Continuous Fins on an Array of Tubes of a Refrigeration Air Cooled Condenser, Int. Journal of Exergy, Vol. 2, No 2, pp 146-171.

Jassim, RK, Mohammed Ali AK, 2003a, "Computer simulation of thermoeconomic optimization of periodic-flow heat exchangers", Journal of Power and Energy, Vol. 217, No.5, pp 559-570

Johnson RS. 1988. The theory and operation of evaporative coolers for industrial gas turbine installations. *GT and Aero-engine Conference*, Amsterdam, Netherlands, 6-9, June, paper # 88-GT-41.

Kakaras E, Doukelis A, Karellas S, 2004. Compressor intake air cooling in gas turbine plants, *Energy*. 29:2347-2358.

Khir, T, Jassim R K, and Zaki, G M, 2007 " Application of Exergoeconomic Techniques to the optimization of a refrigeration evaporator coil with continuous fins", Trans, ASME, Journal of Energy Resources Technology, Vol. 129, No. 3 pp 266-277.

Klein KA, Alvarado FL. 2004. EES-Engineering Equation Solver, Version 6.648 ND, F-Chart Software, Middleton, WI.

Korakiantis T. and Wilson D.G. 1994. 'Models for predicting the performance of Brayton-cycle engines', *Journal of Engineering for gas Turbine and Power*. Vol.116, pp 381-388.

Kotas, TJ, (1995) "The Exergy Method of Thermal Plant Analysis", reprinted, Malabar, Florida, USA: Krieger.

Kotas, TJ, Jassim, RK, Cheung, CF, 1991, "Application of Thermoeconomic Techniques to the Optimisation of a Rotary Regenerator" Int. Journal of Energy Environment and Economics, Vol 1, No. 2., pp 137-145.

Meher-Homji C.B. and Mee R.T. (1999). Gas turbine power augmentation by fogging of inlet air', *Proceedings of the 28th Turbo machinery symposium*. TURBO-Lab: Texas A&M University, Houston, TX, USA. pp 93-117..

Meher-Homji C.B., Mee R.T. and Thomas R. (2002) 'Inlet fogging of gas turbine engines, part B: droplet sizing analysis nozzle types, measurement and testing'. *Proceedings of the ASME Turbo Expo 2002*, Amsterdam, Netherlands, June, paper No GT-30563, V4, 2002, pp 429-441.

Mercer M. (2002) 'One stop shop for inlet cooling systems', *J. Diesel and gas turbine-Worldwide*: V 34 # 6 June 2002, pp. 10-13.

Nag, P.K. (2008) 'Power Plant Engineering', Tata-McGraw Hill. Higher Education, 3rd edition. India.

Ondrays IS, Wilson DA, Kawamoto N, Haub GL. 1991. Options in gas turbine power augmentation using inlet air chilling. *Journal of Engineering for Gas Turbine and Power* 113:203-211.

Ranasinghe, J, Aceves-Saborio, S, Reistad, GM, 1987, "Optimisation of heat exchangers in energy conversion systems", Second Law Analysis of Thermal Systems, proceedings of the forth int. symposium on Second Law Analysis of Thermal Systems, Rome, Italy,29-38 ASME, N.Y.

Rosen MA, Dincer, I, 2003 b. Exergoeconomic analysis of power plants operating on various fuels. *Applied Thermal Engineering*, Vol. 23, No. 6, pp 643-658.

Rosen MA, Dincer, I, 2003 a. Thermoeconomic analysis of power plants: an application to a coal fired electrical generating station. *Energy Conversion and Management*, Vol. 44, No. 17, pp 2743-2761.

Stewart W and Patrick A. 2000. Air temperature depression and potential icing at the inlet of stationary combustion turbines. *ASHRAE Transactions* 106 (part 2):318-327.

Szargut, J., Morris, D.R., and Steward, F.R. (1988) 'Exergy analysis of thermal, chemical and metallurgical process'. Hemisphere, NY, USA.

Tillman TC, Blacklund DW, Penton JD. 2005. Analyzing the potential for condensate carry-over from a gas cooling turbine inlet cooling coil. *ASHRAE transactions* 111 (part 2) DE-05-6-3:555-563.

Tyagi S.K., Chen G.M., Wang Q. and Kaushik, S.C., (2006) 'A new thermoeconomic approach and parametric study of an irreversible regenerative Brayton refrigeration cycle' Int. J. Refrigeration, Vol. 29, No. 7, pp 1167-1174.

Zadpoor AA, Golshan AH, 2006. Performance improvement of a gas turbine cycle using a desiccant-based evaporative cooling system. Energy 31: 2652-2664.

Zaki GM, Jassim, RK, Alhazmy, MM, 2007 "Brayton Refrigeration Cycle for gas turbine inlet air cooling, International Journal of Energy Research, Vol. 31, No. 13 pp 1292-1306.

4

The Selection of Materials for Marine Gas Turbine Engines

I. Gurrappa[1], I. V. S. Yashwanth[2] and A. K. Gogia[1]
[1]Defence Metallurgical Research Laboratory, Kanchanbagh PO, Hydereabad
[2]M.V.S.R. Engineering College, Nadargul, Hyderabad
India

1. Introduction

The desire forever-greater efficiency and increased performance has driven the development in modern gas turbine engines. These engines require high performance materials to exhibit maximum efficiency by increasing their operating temperatures. The blades in modern aero, marine and industrial gas turbines are manufactured exclusively from nickel based superalloys and the compressor section components from titanium based alloys (Fig.1). Achieving enhanced efficiency for marine gas turbines is a major challenge as the surrounding environment is highly aggressive. This aspect depends not only on the design but also on the selection of appropriate materials for their construction. Between the two, selection of materials plays a vital role as the materials have to perform well for the designed period under severe marine environmental conditions. The marine environment makes the superalloys and / or titanium based alloys to undergo a process namely hot corrosion. Hot corrosion can be divided into two types i.e. type I which takes place between

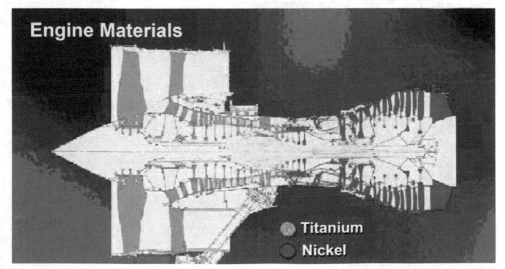

Fig. 1. Significance of superalloys and titanium alloys in gas turbine engines

800 and 950⁰ C and type II that takes place from 600 to 750⁰ C. At higher temperatures, there is no hot corrosion problem as the salt evaporates. Unlike oxidation, hot corrosion is highly detrimental. In fact, hot corrosion is a limiting factor for the life of components for marine gas turbines. Vanadium that is present in the fuel makes the marine environment further corrosive by forming low melting point chemical compounds. Therefore, selection of appropriate materials is paramount importance. An ideal construction material should be able to survive this harsh corrosive environment. Thus, in order to improve the efficiency of marine gas turbine engines significantly, either the existing materials / coatings which can exhibit very good hot corrosion resistance or the advanced materials with considerably improved properties are necessary. Efforts made in this direction made it possible to develop a new superalloy which exhibits excellent high temperature strength properties [1].

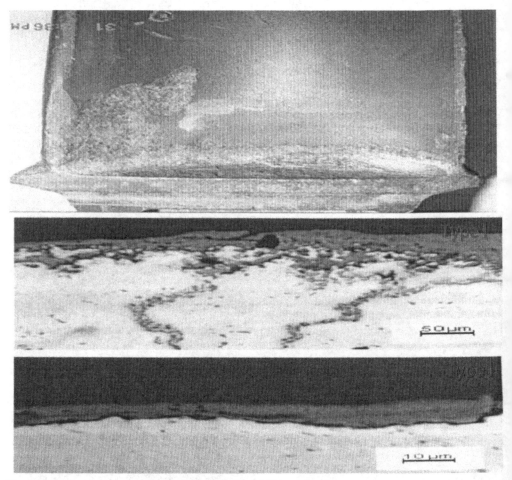

Fig. 2. Failed gas turbine blade due to type I and II hot corrosion

The majority of nickel based superalloy developmental efforts have been directed towards improving the alloy high temperature strength properties with relatively minor concern being shown to its hot corrosion resistance. Further, it is not always possible to achieve both high temperature strength and hot corrosion resistance simultaneously because some alloying elements help to improve hot corrosion resistance while some may help to improve high temperature strength. It is rare that an alloying element leads to enhancement both in high temperature strength and the hot corrosion resistance simultaneously. This is further complicated for marine applications by the aggressivity of the environment, which includes sulphur and sodium from the fuel and various halides contained in seawater. These features are known to drastically reduce the superalloy component life and reliability by consuming the material at an unpredictably rapid rate, thereby reducing the load-carrying capacity and potentially leading to catastrophic failure of components (Fig.2) [2-4]. Thus, the hot corrosion resistance of superalloys is as important as its high temperature strength in marine gas turbine engine applications [5-8]. Recent studies have shown that the high temperature strength materials are most susceptible to hot corrosion and the surface engineering plays a key role in effectively combating the hot corrosion problem [9-13].

2. Superalloys

The selected superalloys for the investigation are presented in Table 1. It is to be noted that SU 263, SU 718, IN 738 LC and IN 792 superalloys contain no rhenium but sufficient amount of chromium. However, SU 263 contains 6% molybdenum and 20% cobalt, iron content is very high in SU 718 with 6% tungsten, 6.5% tantalum and reduced molybdenum 3%. Good amount of tantalum and cobalt 8.5% each and further reduction in molybdenum 1.75% make IN 738 LC. IN 792 contains very low content of tungsten, molybdenum, more amount of aluminium 7.6% and tantalum 5% while CMSX-4 supcralloy has 3% rhenium and reduced chromium. The newly developed alloy contains 6.5% rhenium and a very small amount of chromium. The modified chemistry with 6.5% rhenium, 8.5% tantalum and 5.8% tungsten makes the new superalloy to exhibit very good high temperature strength properties [1].

Superalloy	Ni	Cr	Co	W	Al	Ta	Ti	Mo	Re	Hf	Fe	Mn	Si	Nb
SU-263	Bal	20	20	-	0.6	1.3	2.4	6.0	-	-	0.7	0.6	0.4	-
SU-718	52.5	18.5	9.0	6.0	0.5	6.5	0.9	3.0	-	-	19.0	0.2	0.2	5.1
IN 738 LC	Bal	16	8.5	2.6	3.4	8.5	3.4	1.75	-	-	-	0.2	0.3	0.9
IN 792	Bal	13.5	9.0	1.2	7.6	1.3	5.0	1.2	-	0.2	0.5	-	-	-
CMSX-4	Bal	6.5	9.0	6.0	5.6	6.5	1.0	0.6	3.0	0.1	-	-	-	-
CM 247 LC	Bal	8.1	9.2	8.5	5.6	3.2	0.7	0.5	-	1.4	-	-	-	-
New alloy	Bal	2.9	7.9	5.8	5.6	8.5	-	-	6.5	0.1	-	-	-	-

Table 1. The chemical composition of selected superalloys (wt%)

As hot corroded superalloys like SU 263, SU 718, IN 738 LC in marine and vanadium containing environments under both type II and type I conditions are presented in figures 3-5, while figures 6 and 7 show the hot corroded IN 792, CMSX-4, new superalloy under type II and type I conditions. As can be seen, all the selected superalloys were severely corroded under both the conditions. However, the corrosion is more severe under type I when compared to type II conditions. It indicates that all the superalloys are highly susceptible to hot corrosion. Among them, the new superalloy is more vulnerable to hot corrosion. The new alloy degrades at a very faster rate making it difficult to recognize over a period of time as evidenced from the experiments (Fig.8). It is clearly indicating that the modified chemistry of the new superalloy could not improve its hot corrosion resistance. However, it exhibits very good high temperature strength characteristics as mentioned earlier.

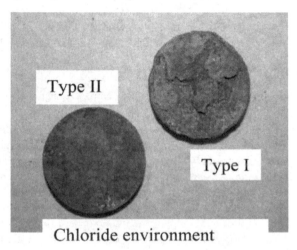

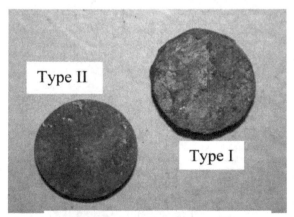

Fig. 3. As hot corroded superalloy SU-263 in marine and vanadium containing environments under type I and type II conditions

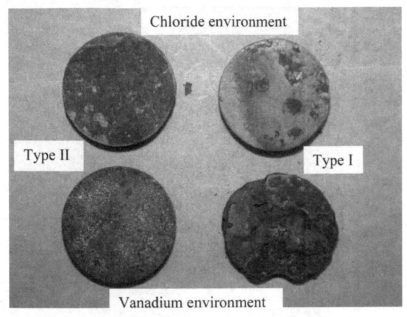

Fig. 4. As hot corroded superalloy SU-718 in marine and vanadium containing environments under type I and type II conditions

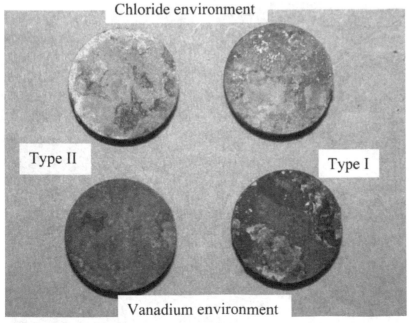

Fig. 5. As hot corroded superalloy IN 738 LC in marine and vanadium containing environments under type I and type II conditions

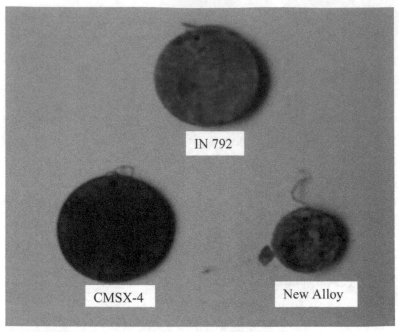

Fig. 6. As hot corroded superalloys in marine environment under type II conditions

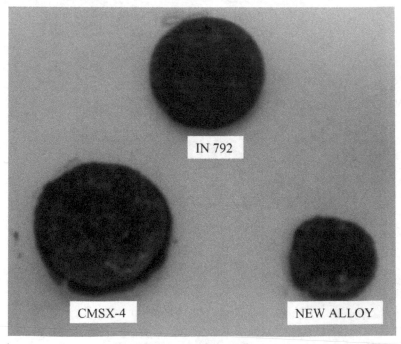

Fig. 7. As hot corroded superalloys under type I conditions in marine environment

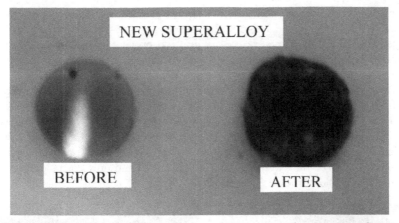

Fig. 8. Before and after hot corrosion in marine environment under type I conditions

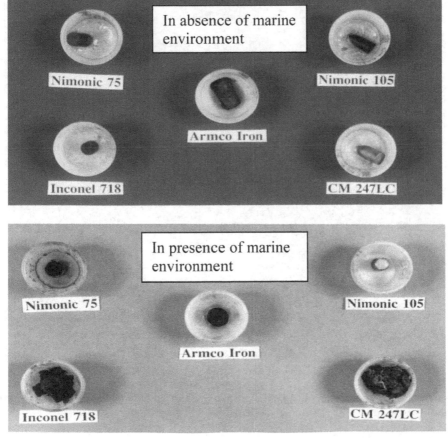

Fig. 9. The effect of marine environment on superalloys under type I conditions

Figure.9 shows the hot corrosion behavior of few more superalloys like Nimonic-75, Nimonic-105, CM 247 LC etc. corroded in the presence and absence of marine environments under type I conditions. In the absence of marine environment, the corrosion was less for all the superalloys [9]. Appreciable corrosion was observed for all the superalloys in the presence of marine environment. It indicates that marine environment plays a significant role in causing severe corrosion, thereby reducing the superalloy life considerably. Among the superalloys, CM 247 LC was corroded severely indicating that this superalloy is highly susceptible to hot corrosion. In fact many cracks were developed on the scale and subsequently spallation took place. However, there were no cracks and no spallation of oxide scales was reported for other superalloys. In case of CM 247LC alloy, no material was left after exposure of 70 hours to the marine environment and only corrosion products with high volume of corrosion products was observed [9].

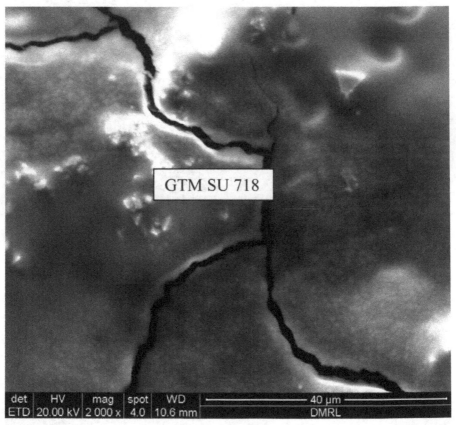

Fig. 10. A typical hot corroded superalloy under type I conditions in marine environment

The surface morphologies of various hot corroded superalloys are revealed that the surface morphology is different for various superalloys under the selected environmental conditions. Electron Dispersive Spectroscopy (EDS) measurements revealed that the corrosion products contain mainly sulphides and oxides of nickel and alloying elements of superalloys like Co, Cr, W, Ti, Ta, Re etc. Typical surface morphology of SU 718 (Fig.10) demonstrates the impact of marine environment by forming big cracks. The cross sections of hot corroded superalloys revealed that the corrosion-affected zone is large for all the superalloys (Fig.11). Among them, the affected zone is more for the new superalloy indicating that severe corrosion took place during the hot corrosion process under marine environmental conditions.

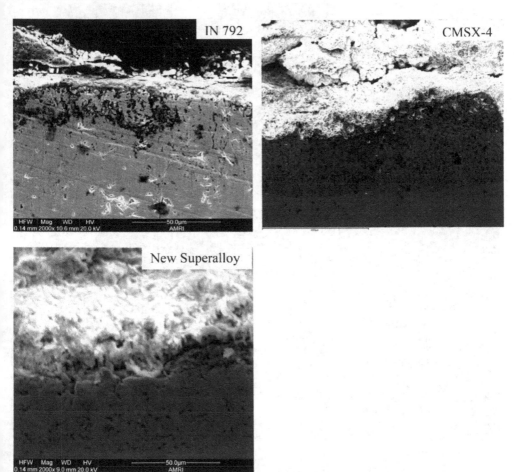

Fig. 11. Cross sections of typical hot corroded superalloys in marine environment

The elemental distributions of hot corroded IN 792 and CMSX-4 superalloys under type I conditions showed that IN 792 superalloy, which contains good amount of chromium (13.5%) could form continuous chromia scale on its surface. It also promoted alumina as well as titania scales. However, extensive diffusion of sulphur and oxygen into the superalloy was clearly observed. While CMSX-4 that contains about 6.5% chromium and 3% rhenium could not form continuous chromia scale. Thin alumina scale was observed on the superalloy surface. Good amount of rhenium was present in the corrosion products. Small amounts of sodium and chlorine were also present in the corrosion products but not diffused into the superalloy. However, significant diffusion of sulphur and oxygen into the superalloy was noticed (Fig.12). Extensive diffusion of sulphur was observed in case of hot corroded CM 247 LC alloy (Fig.13). It is to be noted that neither chlorine nor sodium was diffused into the superalloy.

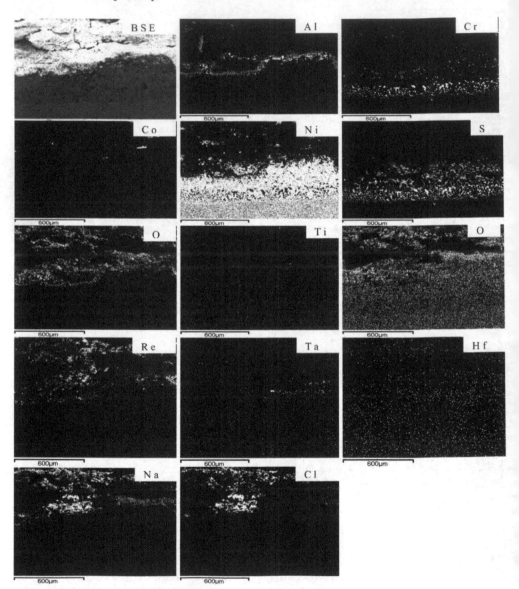

Fig. 12. Elemental distribution of CMSX-4 superalloy after hot corrosion under type I conditions in marine environment

The elemental distribution of hot corroded new superalloy under type I and type II conditions are presented in figures.14 and 15 respectively. The results showed extensive presence of oxygen, sulphur and sodium in the corrosion products. Considerable diffusion of sulphur into the superalloy was clearly observed under type I conditions while oxygen under type II conditions. Rhenium and tungsten were present in the corrosion products under type I and they were present in the corrosion affected zone of new superalloy under type II. Ta and Hf were seen in the corrosion affected region. It is important to mention here that neither alumina nor chromia formation was observed on the superalloy. It is due to the fact that chromium content in the new superalloy is considerably low. At the same time, other alloying elements could not form any protective oxide scales.

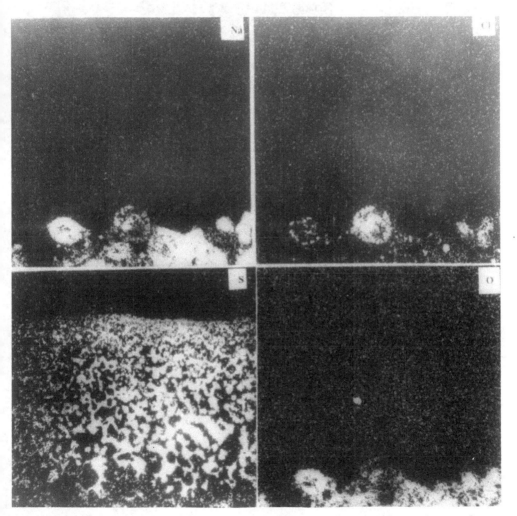

Fig. 13. Elemental distribution of CM 247 LC superalloy after hot corrosion under type I conditions in marine environment

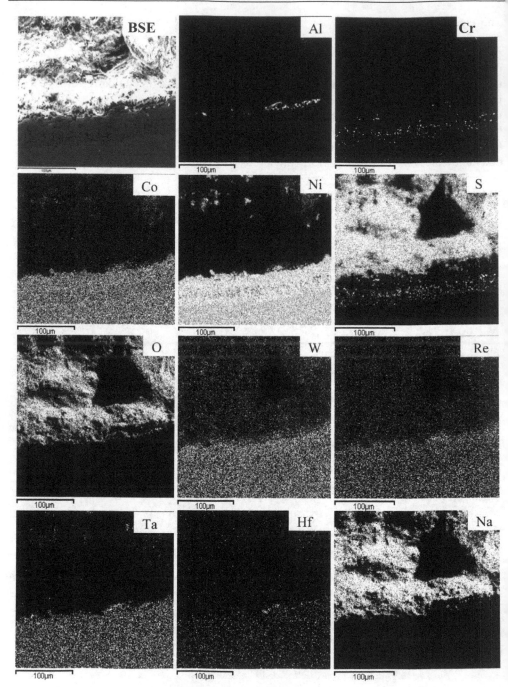

Fig. 14. Elemental distribution of new superalloy after hot corrosion under type I conditions in marine environment

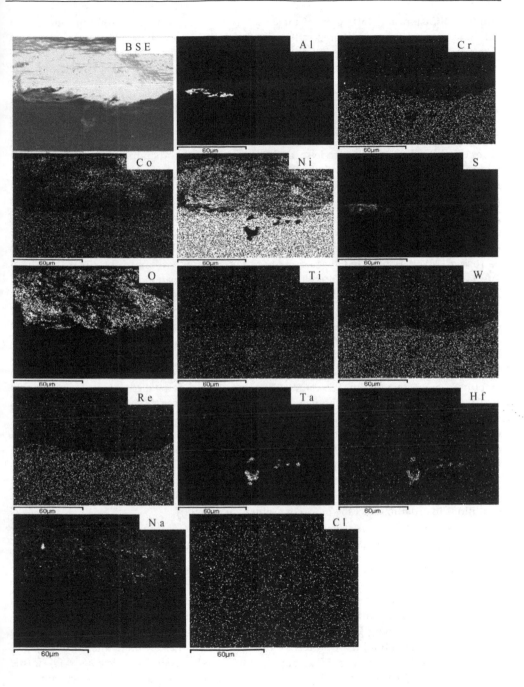

Fig. 15. Elemental distribution of new superalloy after hot corrosion under type II conditions in marine environment

Sulphur diffusion and formation of metal sulphides preferentially chromium and nickel sulphides was reported to be the influential factor. When sulphide phases are formed in superalloys, Ni- based alloys are inferior to cobalt and iron based alloys, which are especially effective in destroying the corrosion resistance of alloys [12]. In essence, the alloying elements play a significant role and decide the life of superalloys under hot corrosion conditions [12].

2.1 Effect of alloying elements

From the metallurgical point of view, it is known that high temperature strength is obtained by maintaining certain phases that are responsible for high temperature strength. Since a main motive for the Metallurgists is to improve the mechanical strength of an alloy at high temperatures, the addition of certain alloying elements is essential with a view to form gamma prime (γ') and solid solution strengthners. Among the alloying elements, the significant reduction of chromium content and the addition of other elements, in particular tungsten, vanadium, molybdenum etc. makes the superalloys more vulnerable to hot corrosion [12]. It is reported that addition of tantalum and titanium produces beneficial effects for hot corrosion [12], while small additions of manganese, silicon, boron and zirconium do not significantly influence the hot corrosion of superalloys. Carbon addition is detrimental to hot corrosion, as the carbide phases provide sites for initiation of hot corrosion [12]. As observed for IN 738 LC with a large amount of chromium and a small amount of titanium, hot corrosion resistance is very good under type I conditions: however the alloy is vulnerable to type II hot corrosion conditions.The addition of molybdenum and large content of iron made the SU 718 less hot corrosion resistant. The addition of large amount of tungsten, tantalum, rhenium and minor other alloying elements and considerably reduced chromium rendered the new superalloy highly susceptible to hot corrosion. It is important to mention that chromium is the most effective alloying element for imporving the hot corrosion resistance of superalloys. In order to obtain good resistance to hot corrosion, a minimum of 15wt% chromium is often needed in nickel based superalloys and a minimum of 25wt% chromium in cobalt based superalloys [4]. However, it is pertinent to note that other alloying elements play a significant role as evidenced from the reported results. Therefore, it is mandatory to test the alloy under simulated environmental conditions in order to select the more corrosion resistant alloy.

2.2 Degradation mechanism

The results clearly revealed that all the studied superalloys are highly vulnerable to hot corrosion. The results further revealed that the new superalloy corrodes much faster when compared to other studied superalloys. It is attributed to the fact that the tungsten which is the alloying element added along with other alloying elements in order to obtain high temperature strength characteristics of the superalloys, forms acidic tungsten oxide (WO_3) due to which fluxing of protective oxide scales such as alumina and chromia takes place very easily. This type of acidic fluxing is self-sustaining because WO_3 forms continuously that cause faster degradation of superalloys under marine environmental conditions at elevated temperatures. The degradation mechanism is explained in two steps as follows:

a)   The tungsten present in the new superalloys reacts with the oxide ions present in the environment and forms tungsten ion

$$WO_3 + O^{2-} = WO_4^{2-}$$

b)   As a result, the oxide ion activity of the environment decreases to a level where acidic fluxing reaction with the protective alumina and chromia can occur

$$Al_2O_3 = Al^{3+} + O^{2-}$$

$$Cr_2O_3 = Cr^{3+} + O^{2-}$$

A similar reaction mechanism occurs if the superalloys contain other refractory elements like vanadium and molybdenum [9].

2.2.1 Electrochemical mechanism

The following section describes an electrochemical phenomenon that explains the new superalloy degradation process in detail under hot corrosion conditions:

Hot corrosion of new superalloy takes place by oxidation of base as well as alloying elements like nickel, cobalt, chromium, aluminium, tantalum, rhenium etc. at the anodic site and forms Ni^{2+}, Co^{3+}, Cr^{3+}, Al^{3+}, Re^{4+}, Ta^{5+}ions etc. while at the cathodic site, SO_4^{2-} reduces to SO_3^{2-} or S or S^{2-} and oxygen to O^{2-}. Since the metal ions i.e. Ni^{2+}, Co^{3+}, Cr^{3+}, Al^{3+}, Re^{4+}, Ta^{5+} ions etc. are unstable at the elevated temperature and therefore reacts with the sulphur ions to form metal sulphides. The metal sulphides can easily undergo oxidation at elevated temperatures and form metal oxides by releasing free sulphur ($MS + 1/2 O_2 = MO + S$). As a result, sulphur concentration increases at the surface of superalloy and enhances sulphur diffusion into it and forms sulphides inside the superalloy. The practical observation of sulphides in hot corroded superalloy specimens clearly indicates that the electrochemical reactions took place during the hot corrosion process. Simultaneously, the metal ions react with oxide ions that are evolved at the cathodic site leading to the formation of metal oxides. The metal oxides dissociate at elevated temperatures to form metal ions and oxide ions. As a result, oxygen concentration increases at the surface and thereby diffuses into the superalloy. Practical observation of oxides in hot corroded superalloys is a clear indication that the electrochemical reactions took place during the hot corrosion process.

Therefore, the hot corrosion of new superalloy is electrochemical in nature and the relevant electrochemical reactions are shown below:

Fig.16 illustrates an electrochemical model showing the new superalloy degradation is electrochemical in nature. Similar mechanism is applicable to other superalloys and their families. The motivation behind suggesting an electrochemical model is to show that the degradation of superalloys in marine environments at elevated temperatures is electrochemical in nature and hence, the electrochemical techniques are quite helpful not only in evaluating them for their hot corrosion resistance but also for understanding their hot corrosion mechanisms. In fact, the electrochemical evaluation of superalloys with and without coatings is more reliable and fast.

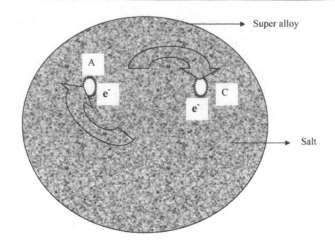

At the anode

$Ni = Ni^{2+} + 2e^-$

$Cr = Cr^{3+} + 3e^-$

$Co = Co^{3+} + 3e^-$

$Al = Al^{3+} + 3e^-$

$Re = Re^{4+} + 4e^-$

$Ta = Ta^{5+} + 5e^-$

etc.

At the cathode

$1/2 O_2 + 2e^- = O^{2-}$

$SO_4^{2-} + 2e^- = SO_3^{2-} + O^{2-}$

$SO_4^{2-} + 6e^- = S + 4O^{2-}$

$SO_4^{2-} + 8e^- = S^{2-} + 4O^{2-}$

Fig. 16. An electrochemical model showing that hot corrosion of new superalloy is an electrochemical phenomenon

2.3 Development of smart coatings

From the present results, it is concluded that the new superalloy is highly susceptible to hot corrosion, though it exhibits excellent high temperature strength properties. It is clear that other superalloys are also vulnerable to both types of hot corrosion. It stresses the need to apply high performance protective coatings for their protection against hot corrosion both at low and high temperatures i.e. type II and type I as the marine gas turbine engines encounter both the problems during service. The protective coatings allow the marine gas turbine engines to operate at varied temperatures and enhance their efficiency by eliminating failures during service. Research in this direction has resulted in design and development of smart coatings which provide effective protection to the superalloy components for the designed period against type I, type II hot corrosion and high temperature oxidation that are normally encountered in gas turbine engines which in turn enhances the efficiency of gas turbine engines considerably [14-15]. This is a major developmental work in the area of gas turbine engines used in aero, marine and industrial applications. Unlike the conventional / existing coatings, the smart coatings provide total protection to the superalloy components used in aero, marine and industrial applications by forming appropriate protective scales like alumina or chromia depending on the surrounding environmental conditions [14-15].

3. Titanium alloys

The titanium alloy components experience hot corrosion problem when they are used for marine gas turbines [16]. It severely limits the high temperature capability of alloys in terms of mechanical properties. It is therefore, desirable to understand the characteristics of titanium alloys under simulated marine gas turbine engine conditions and then apply appropriate coatings, which can prevent hot corrosion and thereby helps in enhancing the life of gas turbine engines significantly.

The hot corrosion characteristics of the titanium alloy, IMI 834 in marine environments at 600^0 C revealed that the rate constant increases by about six times in marine environment and about seven times in vanadium-containing marine environment. It indicates that the rate of reaction is very high in marine environments, still higher in vanadium containing marine environments and low in other environments [16]. The Scanning Electron Micrograph (SEM) of the alloys corroded in marine environment at 600^0 C clearly shows that the oxide scale that formed on the surface of the titanium alloys was cracked due to the presence of NaCl in the environment (Fig.17). The cracks were not observed for the alloys corroded in other environments. It indicates that the chloride ions present in the marine environment causes the oxide scale to crack and facilitates the corrosive species present in the environment to react with the alloy, which is the reason for observing significant increase in corrosion rate [16]. It is known that chloride ions lead to pitting type of attack, which generally initiates at imperfections in the oxide scale. The micro hardness measurements as a function of depth for the alloys corroded at 600^0 C, revealed the presence of about 500 µm hardened zone due to dissolution of oxygen, which is sufficient for affecting the mechanical properties of the titanium alloys by forming a highly brittle zone from which crack initiates during service conditions [16]. The depth of oxygen dissolved

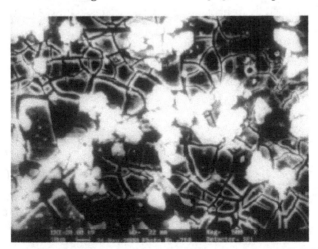

Fig. 17. Effect of marine environment on the stability of titanium alloy IMI 834

region varies with the temperature at which the titanium alloys were corroded. It is important to mention that the depth of the titanium alloys affected in marine environment is about 100 times more than that of the alloys corroded in other environments at the same temperature [16]. It clearly indicates the greater aggressiveness of marine environments to titanium alloys compared to other environments.

3.1 Degradation mechanism

Given below are the proposed mechanistic steps that degrade titanium alloy, IMI 834 under hot corrosion conditions in marine environment:

1. The oxide scale that forms on the surface of IMI 834 is predominantly TiO_2 in association with Al_2O_3. The TiO_2 reacts with chloride ions present in the marine environments at elelvated temperatures to form volatile $TiCl_2$

$$TiO_2 + 2 Cl^- = TiCl_2 + 2 O^{2-} \tag{1}$$

The $TiCl_2$ dissociates at elevated temperatures to form Ti^{2+} and Cl^- ions

$$TiCl_2 = Ti^{2+} + 2 Cl^- \tag{2}$$

The titanium ions then react with oxygen ions present in the environment to form a non-adherent and non-protective TiO_2 scale which spalls very easily. Chloride ions penetrate into the alloy to form volatile chlorides. This process continues until titanium in the alloy is consumed. In other words, the reaction is autocatalytic. The oxygen ions that form in reaction (1) diffuse into the alloy and form an oxygen-dissolution region due to high oxygen solubility in titanium alloys.

2. Al_2O_3 reacts with Cl^- ions to form aluminum chloride

$$Al_2O_3 + 6Cl^- = 2 AlCl_3 + 3 O^{2-} \tag{3}$$

The $AlCl_3$ that formed in the above reaction dissociates to form Al^{3+} and Cl^- ions

$$AlCl_3 = Al^{3+} + 3 Cl^- \tag{4}$$

The Al^{3+} ions react with oxygen ions to form a loose and non-protective alumina scale, which spalls very easily, as in the case of titania

$$Al^{3+} + 3 O^{2-} = Al_2O_3 \tag{5}$$

As mentioned above, the chloride ions penetrate into the titanium alloy to form volatile chlorides and the reaction is autocatalytic. The oxygen ions that formed in reaction (3) diffuse into the alloy and react with titanium. The reactions (1) and (3) contribute to the formation of oxygen dissolved region in the titanium alloy subsurface.

As a result of the above reactions, the degradation of titanium alloys takes place at a faster rate [16] and situation can easily make the components fabricated from titanium alloys, susceptible to failure under normal service conditions of gas turbines. Even in actual jet engines, cracking was reported on salted Ti–6Al–4V alloy discs . Logan *et al* [17] were proposed that oxygen ions from the scale and chloride ions from marine environment, diffuse into the titanium alloys, react with alloys constituents to destroy atomic-binding forces and cause cracking. These detrimental observations clearly stress the need to protect titanium alloy components from hot corrosion and thereby enhance their life by avoiding failures during service. These studies also focus on the development of coatings, which can protect titanium alloys both from oxidation as well as hot corrosion, since both the processes are experienced by gas turbine engine components.

3.2 Smart coatings development

Different smart coatings based on a variety of elements and their combination were designed and developed on titanium alloy, IMI 834. The extensive investigations revealed that the smart coatings based on aluminium that were developed by innovating a new pack composition showed an excellent resistance both under hot corrosion as well as oxidation conditions [15]. The elemental distribution showed a protective, continuous and adherent alumina scale over the coating. It indicates that an excellent protection was provided by the developed smart coating to the titanium alloys from hot corrosion. Further, the developed coatings can be prepared by a simple technique, easy to coat large components and moreover highly economical. Hence, it is recommended to use the developed smart coatings for the modern marine gas turbine engine titanium alloy components.

4. Summary

The chapter presented hot corrosion results of selected nickel based superalloys for marine gas turbine engines both at high and low temperatures that represent type I and type II hot corrosion. The results have been compared with a new alloy under similar conditions in order to understand the characteristics of the selected superalloys. It is observed that the nature and concentration of alloying elements mainly decide the resistance to type I and type II hot corrosion. CM 247LC and the new superalloy are extremely vulnerable to both types of hot corrosion. Relevant reaction mechanisms that are responsible for degradation of various superalloys under marine environmental conditions were discussed. The necessity to apply smart coatings for their protection under high temperature conditions was stressed for the enhanced efficiency as the marine gas turbine engines experience type I and type II hot corrosion during service. Further, the hot corrosion problems experienced by titanium alloy components under marine environmental conditions were explained along with relevant degradation mechanisms and recommended a developed smart coating for their effective protection.

5. References

[1] N. Das, US patent 5,925,198, July 1999
[2] M.R. Khajavi and M.H. Shariat, Engineering Failure Analysis, 11 (2004) 589
[3] J.M. Gallardo, J.A.Rodriguez and E.J. Herrera, Wear, 252 (2002) 264
[4] N. Eliaz, G. Shemesh and R.M. Latanision, Engineering Failure Analysis, 9 (2002) 31
[5] M. Konter and M. Thumann, J. Mater. Process Technol., 117 (2001) 386
[6] J. Stringer, Mater.Sci.Technol., 3 (1987) 482
[7] A.S.Radcliff, Mater. Sci. & Tech., 3 (1987) 554
[8] R.F.Singer, Mater. Sci. & Tech., 3 (1987) 726

[9] I.Gurrappa, Oxid. Met., 50 (1999) 353

[10] C.J. Wang and J.H. Lin, Mater.Chem.Phys., 76 (2002) 123

[11] I.Gurrappa and A. Sambasiva Rao, Surf. Coat. Technol., 201 (2006) 3016

[12] I.Gurrappa, Mater.Sci.Technol., 19 (2003) 178

[13] I.Gurrappa, Surf. Coat. Tech., 139 (2001) 272

[14] Gurrappa, J.Coat. Technol. Res., 5 (2008) 385

[15] Gurrappa, Final Report on "Design and Development of Smart Coatings for Aerospace Applications" submitted to European Commission, July 2008

[16] Gurrappa, Oxid. Met., 59 (2003) 321

[17] H.L.Logan et al., Spec. Tech. Publ. No. 397 (ASTM Materials Park, OH, 1966) p. 215

5

Gas Turbines in Unconventional Applications

Jarosław Milewski, Krzysztof Badyda and Andrzej Miller
Institute of Heat Engineering at Warsaw University of Technology
Poland

1. Introduction

The Chapter presents unconventional gas turbine applications. Firstly some selected non-MAST (Mixed Air and Steam Turbine) solutions are discussed – these are intended for smaller gas turbine systems, where the regenerative heat exchanger supplies energy for an additional thermal cycle, where it is utilised.

The gas turbine cycle (Brayton) may be coupled with several other thermal engines (like another Brayton, Diesel, Kalina, and Stirling). Those hybrid systems have several previously unrecognised advantages. They may find applications in some market niches.

The next section describes hydrogen-fuelled gas turbine solution. Big international programme – WE-NET – is discussed. Several hydrogen-fuelled gas turbine concepts based on those programmes are proposed: Westinghouse, Toshiba, Graz, New Rankine. The section provides description of them all, including specification of possible efficiency values. The development programmes themselves are also reviewed. This part of the text describes also potential combination of a hydrogen-fuelled gas turbine and a nuclear power generation unit which might be used to cover peak load power demands in a power system.

Last section of the chapter discusses integration of a fuel cell into a gas turbine system. High temperature fuel cells can play a role similar to a combustion chamber but simultaneously generating additional power. Fuel cell hybrid systems for both high-temperature types of fuel cells – Solid Oxide Fuel Cell (SOFC) and Molten Carbonate Fuel Cell (MCFC) – are proposed. Additionally, some specific properties of the MCFC can be used to reduce carbon dioxide emissions from the gas turbine itself.

Gas turbine systems, particularly combined cycle units, are among the most popular power systems in the modern world. This results from the very fast technical progress allowing to gradually increase the parameters at the turbine inlet as well as unit outputs. There is also a parallel development trend of searching for new unconventional solutions, which would allow to achieve efficiencies higher than enabled by a simple cycle. Scheme of the simple cycle process is presented in Fig. 1. Simple cycle efficiency in many cases is too low.

Internal power N_i of a gas turbine can be obtained by an analytical approach by using the relation 1, which is obtained with assumption that the process is real (contained losses) and working fluid is modelled as semi-ideal gas.

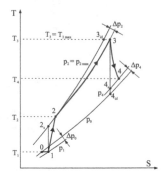

Fig. 1. Thermal process of a simple cycle gas turbine

$$N_i = N_T - N_K$$

$$= G_T \cdot c_{p,T} \cdot T_3 \left(1 - x_T\right) \cdot \eta_T - G_K \cdot c_{p,K} \cdot T_1 \cdot \left(x_K - 1\right) \cdot \frac{1}{\eta_K} s \tag{1}$$

$$= G_K \cdot c_{p,K} \cdot T_1 \cdot \left[\frac{1}{1 - \beta} \cdot \bar{c}_p \cdot \Theta \cdot \left(1 - x_T\right) \cdot \eta_T - \left(x_K - 1\right) \cdot \frac{1}{\eta_K}\right]$$

where: N_T, G_T, N_K, G_K stand for internal power and fluid inlet mass flow in turbine and compressor, respectively, η_T – turbine internal efficiency, η_K – compressor polytropic efficiency, $c_{p,T}$, $c_{p,K}$ – averaged isobaric specific heat capacity for the working fluid in turbine and compressor, k_K, k_T – averaged heat capacity ratios (isentropic exponents) for compressor and turbine for the flue gas and air. Other symbols used in Eq. 1 are:

$$\bar{c}_p = \frac{c_{p,T}}{c_{p,K}}; \quad \Theta = \frac{T_3}{T_1}$$

$$x_T = \frac{1}{\Pi_T^{m_T}}; \; m_T = \frac{k_T - 1}{k_T}; \; \Pi_T = \frac{p_3}{p_4}$$

$$x_K = \Pi_K^{m_K}; \; m_K = \frac{k_K - 1}{k_K}; \; \Pi_K = \frac{p_2}{p_1} = \Pi \tag{2}$$

$$\Pi = \Pi_K = \frac{p_2}{p_1}; \quad \Pi_T = \frac{p_3}{p_4} \quad \Pi_T = \varepsilon \Pi_K$$

$$\beta = \frac{G_p - \Delta G}{G_T}$$

where: G_p – fuel mass flow delivered to a combustion chamber, ΔG – air mass flow delivered for cooling purposes of the hottest parts of gas turbine and bearings, and leakages losses, Π_T, Π_K are pressure ratios of outlet and inlet of compressor and turbine, and losses factor ε describes combination of pressure losses at compressor inlet, combustion chamber and turbine outlet.

Subscripts of working fluid parameters (pressure and temperature) indicate location in reference to the turbine diagram (Fig. 2). Specific internal power by definition is expressed as:

$$N_j = \frac{N_i}{G_K} \tag{3}$$

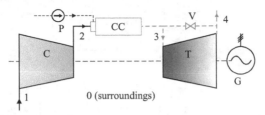

Fig. 2. Simple cycle gas turbine diagram: C – compressor, T – turbine, G – electric generator, CC – combustion chamber, P – fuel pump, V – bypass valve

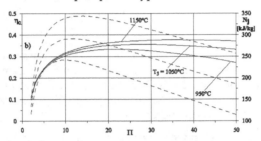

Fig. 3. Specific power output and thermal efficiency of a simple cycle gas turbine unit as functions of the pressure ratio. Dashed lines denote specific power output.

and thermal efficiency is a ratio of internal power and fuel power input:

$$\eta_c = \frac{N_i}{N_p} \qquad (4)$$

Characteristics of a simple cycle gas turbine unit based on the relationships presented above are shown below. Performance calculations were carried out at variable pressure ratio Π_K and temperature T_3 (at turbine inlet, without taking into consideration cooling losses, i.e. $\Delta G = 0$). Constant values of polytropic efficiencies of turbine $\eta_T = 0.88$ and compressor $\eta_K = 0.88$, pressure losses factor $\varepsilon = 0.95$ were used. Ambient conditions (pressure and temperature) were assumed according the ISO conditions: $T_0 = 288$ K (15°C).

The first curve (Fig. 3) illustrates key performance parameters of a simple cycle gas turbine unit (specific power output and efficiency) as functions of pressure ratio and Turbine Inlet Temperature (TIT). It can be seen that in a real system maximal specific power output is achieved at lower pressure ratio then the maximal efficiency.

Another curve shows relation between efficiency and a specific power output (Fig. 4). This curve was obtained with the same relationships and assumptions as for Fig. 3. The range of investigated temperatures was extended (while keeping the assumption that there is no turbine cooling, which results with performance figures being somewhat exaggerated). Points on individual curves denote selected pressure ratio values. Analysis of the cooling system impact on performance of a gas turbine can be found for example in Nat (2006).

It is possible to considerably increase efficiency and other performance figures of a gas turbine unit at the expense of introducing new components and making the flow system more complex. Therefore, except for simple gas turbines, also complex units with more

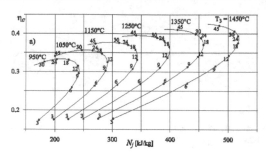

Fig. 4. Thermal efficiency of a simple cycle gas turbine unit as a function of specific power and turbine inlet temperature.

sophisticated working cycle are developed and build. These may involve for example heat regeneration, inter-stage cooling and re-heating. Such solutions require introduction of new components: regenerative (or recuperative) heat exchanger, gas coolers, combustion chamber cooling systems or inter-stage heaters. Also the compressor and turbine need to be split into high- and low-pressure parts (or even more sections).

In parallel to development of simple cycle gas turbines, also combined cycle solutions are developed, among them units with additional thermal circuit. Technical progress in gas turbine development is coupled with development of combined cycle systems, including solutions with combination of different thermal engine types. Combination of a gas turbine thermal cycle with an intermediate- or low-temeprature steam power plant cycle may be seen as the most efficient method to increase the efficiency of a process involving gas turbines used in industrial practice. Nowadays gas turbine combined cycle solutions are among the most intensely developed power generation technologies. "Conventional" Gas Turbine Combined Cycle or GTCC solutions may reach a nominal efficiency reaching 60%.

It is however not possible to achieve such high efficiencies at intermediate output GTCC units. It may prove more feasible to utilise simpler systems at lower scales. Specific investment cost for low-output GTCC unit is quite high, while the efficiency may be considerably lower. They do display high reliability though. One method to improve performance of low-output GTCCs may be utilising recuperation, inter-stage compressor cooling or working agent integration, e.g. utilising air-steam mixture as a working agent.

This last concept utilised in gas turbine systems has been proposed in various forms and under different names. One of the used names is Mixed Air and Steam Turbine (MAST). Another name is Cheng Cycle, and General Electric utilises name Steam Injected Gas Turbine – STIG. The gas turbine concepts with air humidification are known as Humid Air Turbines (HAT), Cascaded Humidified Advance Turbines (CHAT) or Wet Compression systems.

2. Gas turbines with other thermal engines

2.1 Brayton-Brayton

The Brayton-Brayton cycle (diagram shown in Fig. 5) is a combination of two simple-cycle gas turbines. Working agent in one of them is flue gas, and in the other – air. The turbines are interconnected with a high-temeprature air/flue gas heat exchanger. As the air turbine utilises exhaust heat, this solution enables to considerably improve total efficiency when compared to

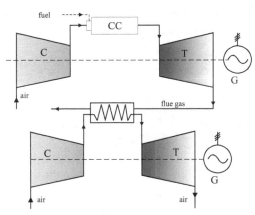

Fig. 5. Diagram of a combined cycle composed of a gas turbine and an air turbine supplied with heat recovered from the flue gas flow. (Brayton-Brayton)

a simple cycle solution. When compared to a classic GTCC, the Brayton-Brayton techenology requires less auxiliary equipment. It also needs less space and has lower investment cost. Systems of this type are not extensively analysed in the literature. Discussion presented in this chapter is based on the analyses presented in detail in Baader (n.d.). Performance of each gas turbine utilised in the system may be determined according to the rules and relations presented in the introduction.

Performance of the entire system depends on parameters (selection criteria) of the heat exchanger and air-based turbine unit. Calculations carried out in order to determine performance of a system shown in Fig. 5 were based on the following assumptions:

1. The air cycle is designed to maximally utilise energy carried with the flue gas exhausted from the gas turbine unit. The flue gas is cooled as far as possible, but to temperatures no lower than 200°C. Efficiency of the flue/air heat exchanger is 80% and the pinch point temperature is 30°C.

2. Pressure ratio of the air turbine is designed to enable achieving highest possible internal power.

3. Assumed compressor and turbine polytropic efficiencies are 88% (just like for the simple cycle).

4. Assumed pressure losses at the heat exchanger for both air and flue gas is 3.4%, which results with the pressure loss coefficient of 0.928.

5. Just like in case of previous calculations no impact of internal gas turbine cooling systems on system's performance is taken into account.

Other assumptions and the relationships used followed the rules previously presented for the simple cycle. Results are shown in the charts – characteristic curves for the Brayton-Brayton system. Fig. 7 shows relation between the system's efficiency and specific power output. Values shown here indicate that a considerable increase in reference to a simple cycle system may be expected (compare to Fig. 2).

Maximum specific power occurs at a lower pressure ratio value than for a simple cycle system, but at higher values than for a system with regeneration. Specific power is much higher than in case of a simple cycle or a regenerative system.

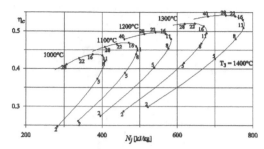

Fig. 6. Thermal efficiency of a Brayton-Brayton system as a function of specific power output and Turbine Inlet Temperature. Numerical values shown in the chart denote the gas turbine pressure ratio.

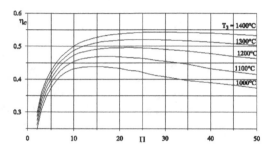

Fig. 7. Thermal efficiency of a Brayton-Brayton system as a function of gas cycle's pressure ratio and turbine inlet temperature

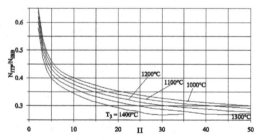

Fig. 8. Share of the internal power generated in the air turbine in a total internal power of a Brayton-Brayton system as a function of gas cycle's pressure ratio and turbine inlet temperature.

Fig. 7 shows relation between the efficiency of a Brayton-Brayton system and the pressure ratio in the gas cycle. The next figure (Fig. 8) illustrates distribution of internal power between the gas and air cycles. High share of the air cycle at low gas cycle pressure ratio values results from the high temperature of exhaust gas delivered to the heat exchanger downstream from the gas turbine unit in such a case. This allows to achieve high pressure ratio in the air cycle – as already mentioned this value is optimised to achieve highest possible output. As shown in Fig. 9 the pressure ratio of the air cycle at low pressure ratio of gas cycle is high Π_{KP}/Π much lower than one). The results in this area should be seen as exaggerated, as the model does not include any limit of maximum (achievable) air temperature downstream from the heat

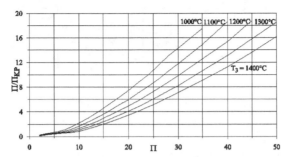

Fig. 9. Ratio between the gas cycle pressure ratio and air cycle pressure ratio in a Brayton-Brayton system as a function of the gas cycle pressure ratio.

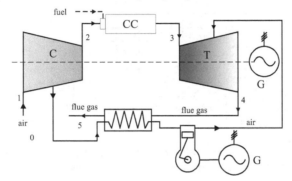

Fig. 10. Brayton-Diesel system flowchart

exchanger. In fact such a restriction would result from the material properties – this effect would practically rule out systems with low gas cycle pressure ratio. Practically only systems where the gas cycle parameters are similar to those used in simple cycle systems, i.e. those close to the parameters range enabling maximum specific power outputs and efficiencies (see Fig. 3 and Fig. 4), should be seen as reasonable.

2.2 Brayton-Diesel

One of the interesting proposals presented in the literature is the so-called Brayton-Diesel system Poullikkas (2005).

The Brayton-Diesel system (Fig. 10) is a combination of a simple cycle gas turbine with a heat exchanger and a reciprocating expander. Working agents are exhaust gas and air. Part of the air flow from the compressor is supplied to the combustion chamber, while some air is transferred into a heat exchanger where it is heated by the gas turbine exhaust. Then this flow expands in a reciprocating expander and is further fed into the low-pressure stages of the gas turbine, mixing with the exhaust gases. It was assumed that the mixing process is isobaric. The resulting mixture further expands to the turbine exhaust pressure.

Application of the Brayton-Diesel system allows to increase maximum power output of a single unit by some 6–12% and increase maximum efficiency by 1.3–3.6% in the investigated temperature range (900–1,300°C).

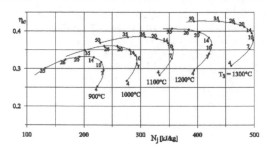

Fig. 11. Thermal efficiency of a combination of a gas turbine and reciprocating expander (configuration as in Fig. 10) as a function of specific power and TIT (T_3). Numerical values shown in the chart denote gas turbine pressure ratios.

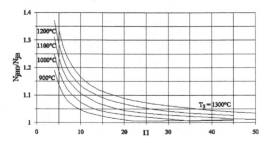

Fig. 12. Increase of a specific power output of a system with a reciprocating expander in comparison to the specific power output of a simple cycle gas turbine as a function of gas turbine pressure ratio and TIT (T_3).

An interesting solution could be also supplying fuel to the reciprocating expander, which would also require combustion process there. This case is currently a subject of further investigation.

The Brayton-Diesel system can also be used for combined heat and power applications. Mixture of exhaust gas and air at the outlet of the heat exchanger has a temperature in range of 170–330°C, so it can be used as a source of process heat.

Internal power of the combined cycle N_{iz} was calculated according to the rules and relationships given in Miller (1984), by binding it to the parameters of the thermal process according to equations:

$$N_{iz} = N_{Tz} + N_R - N_{Kz} \qquad (5)$$

where: N_{Tz}, N_R, N_{Kz} – internal power of the turbine, reciprocating expander and combined-cycle compressor, respectively.

Specific power of the system was defined as follows:

$$N_{jz} = \frac{N_{iz}}{G_K} \qquad (6)$$

Analysis of the results obtained for the Brayton-Diesel system model allows to draw following conclusions:

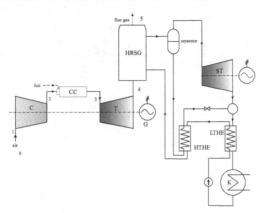

Fig. 13. Process diagram of a Brayton-Kalina system: C – compressor, T – turbine, G – generator, CC – combustion chamber, ST – steam turbine, K – Condenser, HRSG – heat recovery steam generator, HTHE – high temperature heat exchanger, LTHE – low temperature heat exchanger.

1. Combination of a simple-cycle gas turbine with a heat exchanger and a reciprocating expander increases thermal efficiency and specific power of the system.

2. Maximal thermal efficiency and maximal specific power increase; the higher the maximum temperature in the system – the higher growth of those parameters.

3. Maximal thermal efficiency grows slightly, the growth of the specific power is more significant.

4. Optimum values of the simple cycle pressure ratio are lower for the combined cycle than for the simple cycle.

5. Using reciprocating expander is most profitable for possibly high maximal temperature of the system and possibly low pressure ratio of the simple cycle.

6. Heat exchanger used in this system has to provide significant amount of heat to the air (ΔT_{pow} = 173...470°C) so it needs to have quite large heat exchange area which makes it expensive and bulky.

7. Discussed system can also be used in combined heat and power applications as the air-exhaust gas mixture flowing out of the heat exchanger has a quite temperature in range 170...330°C.

While selecting design parameters of the Brayton-Diesel system also the economical analysis is very important. Increasing TIT and using very large heat exchanger could prove too expensive when compared to the benefits resulting from the higher efficiency.

2.3 Brayton-Kalina

The Kalina system is a variant of the Organic Rankine Cycle (ORC). It utilises a mixture of water and ammonia as a working agent. Ratio between ammonia and water may be changed, depending on the process, which enables adjusting boiling and condensation temperatures. The Kalina cycle is based on the Rankine cycle with added distilling and absorption components (Fig. 14). Its main advantage is already mentioned ability to adjust boiling and

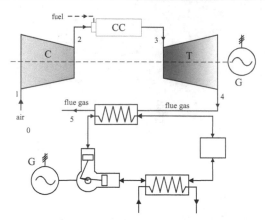

Fig. 14. Process diagram of a Brayton-Stirling system. C – compressor, T – turbine, G – generator, CC – combustion chamber

condensation temperatures of the working agent, allowing to adjust entire cycle to the variable temperature of the heat source. Changing boiling and condensation temperatures during plant operation provide an additional degree of freedom when compared to a traditional steam cycle (including ORC solutions). Variable water content in the ammonia solution allows to optimise operational parameters of the cycle and thus improve the efficiency which may be close to the Carnot cycle.

Working agent properties allow to decrease the temperature difference in reference to the exhaust gas temperature in the heat recovery steam generator (the pinch point effect in evaporator is missing).

The Kalina cycle has a number of features making it competitive against the Rankine cycle (both "classic" and organic):

1. It may produce 10–30% energy more than a Rankine cycle.
2. It has lower (by some 40%) space requirements than a corresponding system with a steam turbine.
3. Exhaust steam pressure is higher than ambient pressure so it is not required to create vacuum in the condenser – this enables to shorten the start-up time.
4. The Kalina cycle can be easily optimised for changed ambient conditions by adjusting working agent composition.

The licence for construction of Brayton-Kalina systems (process illustrated in Fig. 14) has been owned by General Electric since 1993. While further technology development was declared and a commercial power plant based on it was supposed to be completed by 1998, in fact the programme was suspended.

2.4 Brayton-Stirling

The combination of Brayton and Stirling cycles can have different configurations. Heat to the Stirling engine may be delivered in two different points of the cycle. One possibility is heat transfer through a heat exchanger installed directly inside the combustion chamber of a

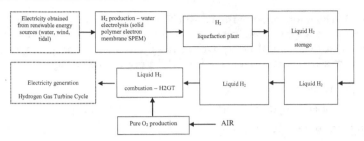

Fig. 15. Configuration of a hydrogen-based power generation system

gas turbine unit. The other possibility is recovering exhaust gas heat downstream from the gas turbine (as shown in the Fig. 14). Performance of this configuration may be optimised, with one of the variables being materials used for construction of the main heat exchanger. A Brayton-Stirling plant is able to achieve relatively high efficiency values. In a system built some years ago and utilising a Rolls-Royce RB211 gas turbine installation of a Stirling engine allowed to increase the power output by 9 MW (from the original 27.5 MW). This increased the total efficiency Poullikkas (2005) to 47.7%. And advantage of such a solution is its compact form and simplicity of waste heat utilisation.

3. Hydrogen gas turbine

The issue of environment pollution caused by human social and economical activities in recent years, has caused an increase of interest in environment-friendly (zero-emission) technologies of energy generation and utilisation. This approach is commonly seen as a way to solve global environmental problems that have recently become more and more intense. Environmental footprint of classic energy generation technologies based on fossil fuels – which are currently in use – is an argument for developing alternative technologies for energy generation. Power generation based on release of chemical energy of combusted hydrocarbon fuels is one of the main causes of the biosphere degradation. Combustion of fossil fuels (coal, lignite, oil, gas) causes an environmental footprint: emission of combustion to the atmosphere or the problem of combustion waste management which still remains unsolved.

An increase of CO_2 concentration in the atmosphere intensifies the greenhouse effect and thus influences climate changes. Moreover, SO_x and NO_x emissions caused by many industry processes, especially combustion used in energy industry, causes acid rains falling in industrial areas.

Hydrogen as a clean fuel is a subject of interest for many research institutions all over the world. National and international research projects aimed at utilising hydrogen generated by renewable energy sources (e.g. wind or solar) have been proposed in various parts of the world Gretz (1995); Kaya (1995).

3.1 Review of the WE-NET system concept with hydrogen turbine

The Japanese World Clean Energy Network programme using hydrogen conversion (WE-NET) started in 1993 as a part of larger "New Sunshine Program", whose main goal was to develop environment-friendly energy generation technologies that could meet constantly increasing energy demand and at the same time allow to solve arising environmental problems (n.d.). The main target of the WE-NET project is to construct a worldwide energy network for effective supply, transmission and utilisation of non-carbonaceous renewable energy (water, solar, wind, sea) using hydrogen as a clean secondary energy carrier. The energy conversion system proposed in the WE-NET programme includes:

- Hydrogen production through water electrolysis

- Conversion of gaseous hydrogen into forms enabling easy transport

- Storage and maritime transport, and

- Utilisation in multiple areas (including power generation by hydrogen combustion turbine systems).

Diagram of such a power generation system is shown in the Fig. 15.

Alternative energy generation technologies based on hydrogen and proposed by the WE-NET programme could limit emissions of CO_2 and other greenhouse gases to the atmosphere. Additionally this technology would enable using renewable energy sources on large industrial scale. It is due to the fact that H_2 would be produced from water by electrolysis process powered by renewable energy. Present feasibility of the utilisation of renewable energy sources depends on a specific project site and scale, ergo depends on a local energy demand and is also limited by the losses in electricity transformation and distribution processes.

Last but not least, CO_2 emission has recently become one of the most important subjects of the environment protection. In order to meet the newly defined environmental requirements new and more efficient technologies need to be researched. Performance of an ordinary GTCC unit is approaching the maximum efficiency level limited by NO_x emission problem. Thus a Hydrogen Combustion Turbine System could be seen as an interesting research subject, because it does not cause any NO_x or CO_2 emissions and could be able to achieve over 60% (HHV) efficiency.

Various cycles for the hydrogen combustion turbine system could be proposed to obtain high thermal efficiency (over 60% HHV). 500 MW class gas turbine system with turbine inlet temperature of 1,700°C was proposed as a result of the preliminary research. High performance parameters are crucial, as hydrogen is still a very expensive fuel (due to high cost of production, storage and transport).

The Japanese WE-NET Program NEDO (1997) predicts an implementation of the Hydrogen-Fuelled Combustion Turbine Cycle (HFCTC) as a new energy source for power generation sector. In this respect, a configuration and performance study of the HFCTC was conducted Miller et al. (2001). The WE-NET Program aims at establishing a global energy network using renewable energy based on utilisation of hydrogen – a secondary clean energy carrier. Production cost of hydrogen is relatively high, so it is required to design a power generation system with thermal efficiency superior to that of existing conventional power units. Thermal efficiency above 60% HHV is required (it means above 71% LHV).

Nowadays, thermal efficiency of the most advanced combined cycles (natural gas-fuelled) is close to 60% LHV (54% HHV) and it is comparable to 50.4% HHV if hydrogen fuel is used.

It is obvious that efficiency should be increased by about 10 percentage points, which is an equivalent of about 20% in comparison to the most efficient contemporary power plant units. It is a very serious technological challenge (a qualitative change, not only quantitative).

The development target was to build a full-scale 500 MW_e power generation unit. Besides increasing working agent's temperature at the turbine inlet to 1,700°C (at present it is a level of 1,500°C), it is essential to implement a new approach (different than traditional one) to both conceptual design of the system (its configuration and working parameters) and detailed design solutions. The issue was challenging and potentially beneficial.

Taking the above into account, the main problem was to identify correct cycle concepts and evaluate their performance characteristics. Due to its advantageous features, the GRAZ cycle was treated as a basic concept in the WE-NET Program Mouri (1999). However, during the research conducted by the authors several other concepts were analysed Iki et al. (1999); Miller et al. (2002). The research included, among others, the following main items:

- Cycle identification and assessment,
- Cycle selection,
- Comparison of selected cycles' performance in nominal state under the same reference conditions.

Until now several concepts of the HFCTC have been proposed, out of which the most important are:

1. Combined Steam Cycle with Steam Recirculation developed by Prof. H. Jericha (Technical University of Graz), commonly referred to as the GRAZ cycle Iki et al. (1999); Jericha (1984); Moritsuka & Koda (1999).
2. Direct-Fired Rankine Steam Cycle (New Rankine Cycle) which was studied in the following variants:
   (a) proposed by Toshiba Co., commonly referred to as the TOSHIBA cycle Funatsu et al. (1999); Moritsuka & Koda (1999);
   (b) proposed by Westinghouse Electric Co., commonly referred to as the WESTINGHOUSE cycle Bannister et al. (1997);
   (c) Modified New Rankine Cycle, as authors' own concept, commonly referred to as the MNRC cycle.

A common feature of the abovementioned cycles is that only one working medium (steam) is used for both the topping and bottoming cycles. This is enabled by replacing external firing system (as in the Rankine steam cycle) with direct firing concept (similar to gas turbines or reciprocating engines). Main assumption made here is stoichiometric combustion of hydrogen and oxygen mixture. This combustion takes place inside a stream of a cooling steam, which cuts down combustion temperature to 1,700°C. It is also assumed that hydrogen and oxygen are available at the ambient temperature and at pressure level that allows to inject them into a combustor. It means that hydrogen would need to be provided as cryogenic liquid, however cryogenic energy could be utilised for pure oxygen production in an air-separator unit.

GRAZ, TOSHIBA, WESTINGHOUSE and MNRC cycles were analysed in comparable conditions to evaluate their performance. The analysis was undertaken for the same specific conditions with the same assumptions and property tables. Results of the research for abovementioned cycles published so far, do not provide an opportunity for such a comparison because of incomparable and/or not clear conditions and assumptions of individual studies. Cycles without cooling system were only taken into account in order to create an opportunity for an explicit evaluation.

Fuel type	η_{LHV}/η_{HHV}
Hard coal	1.05
Natural gas	1.11
Hydrogen	1.19

Table 1. Relation between LHV and HHV efficiency for different types of fuel.

3.2 Performance analysis of WE-NET systems with hydrogen turbine

3.2.1 Theory

At the beginning of this section some remarks on cycles' efficiency definition are presented. There are four kinds of efficiency definition used for the comparative analysis:

1. Carnot efficiency which defines theoretical limits for the efficiency of the specific cycle:

$$\eta_C = \left(1 - \frac{T_{bottoming}}{T_{topping}}\right) \cdot 100\% \tag{7}$$

where: T – absolute temperature, K.

2. Rankine cycle efficiency related to the simple steam Rankine cycle:

$$\eta_R = \frac{P}{Q_{in}} \cdot 100\% = \frac{P}{P + Q_{cond}} \cdot 100\% = \frac{1}{1 + \frac{Q_{cond}}{P}} \cdot 100\% \tag{8}$$

where: P – output power of the cycle, MW, Q – heat flow, in – input, $cond$ – output.

3. HHV efficiency related to higher heating value:

$$\eta_{HHV} = \frac{P}{m_{fuel} \cdot HHV} \cdot 100\% \tag{9}$$

where: m – mass flow, kg/s; HHV – Higher Heating Value of the fuel, MJ/kg.

4. LHV efficiency related to lower heating value:

$$\eta_{LHV} = \frac{P}{m_{fuel} \cdot LHV} \cdot 100\% \tag{10}$$

where: m – mass flow, kg/s; LHV – Lower Heating Value of the fuel, MJ/kg.

HHV is LHV plus heat of evaporation of water content in the fuel (2,500 kJ/kg). The larger amount of hydrogen in the fuel is, the higher the difference between HHV and LHV efficiencies gets. If coal is the fuel then the heat from hydrogen content is practically impossible to recover. This is why LHV is commonly used for thermodynamic calculations of coal fired cycles. But when the fuel is the pure hydrogen, then steam content in combustion gases is 100% and HHV is more convenient to use.

The relative difference between η_{LHV} and η_{HHV} is about 20%. Relation between LHV and HHV efficiency is shown in the Table 1.

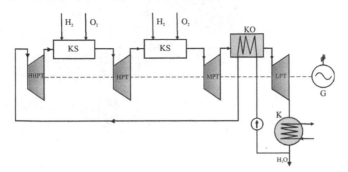

Fig. 16. TOSHIBA' cycle diagram

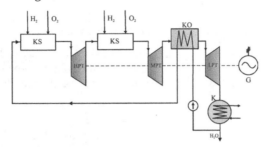

Fig. 17. WESTINGHOUSE cycle diagram

3.2.2 Toshiba cycle

Toshiba Co. proposed the cycle as the company's contribution to the WE-NET project Funatsu et al. (1999). The concept is based on standard Rankine cycle. The regeneration heat exchanger replaced the boiler of the Rankine cycle. Reheating is carried out by two combustors with H_2-O_2 combustion in a steam flow. An implementation of four turbine parts and two combustors is characteristic for the TOSHIBA cycle. The TOSHIBA cycle thermal diagram is shown on the Fig. 16.

3.2.3 Westinghouse cycle

Westinghouse Electric Co. proposed the cycle as the company's contribution to the WE-NET project Bannister et al. (1998). The concept is based on standard Rankine cycle. The regenerative heat exchanger and one H_2-O_2 combustor replace the boiler. Re-heat is performed via one H_2-O_2 combustor. The WESTINGHOUSE cycle consists of two H_2-O_2 combustors and three turbine parts. The WESTINGHOUSE cycle thermal diagram is shown in the Fig. 17.

3.2.4 The GRAZ cycle

A combined steam cycle with steam recirculation, proposed by prof. Herbert Jericha from Graz University of Technology (Austria) Jericha (1991) is called the GRAZ cycle. Graz cycle consists of one combustor, three turbine parts, heat recovery steam generator, condensing part and compressor. The GRAZ cycle thermal diagram is shown in the Fig. 18. The GRAZ cycle is an interesting composition of Brayton and Rankine systems. The Brayton cycle is used as

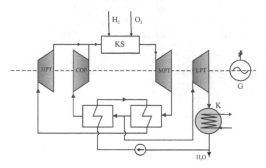

Fig. 18. Flow diagram and calculation results for the GRAZ cycle

a topping cycle (high parameters zone) in semi-closed arrangement, and is coupled to the Rankine cycle operating as the bottoming cycle (low parameters zone).

Basically, the cycle consists of one combustor, three turbine parts, heat recovery steam generator, condensing part and compressor. Thermal efficiency achieved in the cycle amounts to 61% HHV (version without cooling). A flow diagram and basic flow parameters of the GRAZ cycle are shown in Fig. 18.

As in other cycle concepts of this type, the single working medium (steam) is used in both topping and bottoming cycle. Replacement of an external firing (as in the Rankine steam cycle) by a direct firing (similar to gas turbines or reciprocating engines) is also a characteristic feature. The combustion takes place inside a cooling steam flow, which reduces combustion temperature to 1,700°C. Combustion process of hydrogen-oxygen mixture was assumed to be stoichiometric.

An original concept, which distinguishes the GRAZ cycle from others, was a bleed of partially cooled working fluid from the Brayton cycle (point 8 of flow diagram, Fig. 18) and its utilisation as working fluid in the Rankine cycle (points 8, 9, 12, Fig. 18). Thus, this cycle is sometimes called the topping-extraction cycle. Efficiency benefit in this case comes from substantial decrease of compression load in the Brayton cycle, because large share of working fluid (more than a half) reaches high pressure in a feed water pump (points 12, 10: Fig. 18), instead of being compressed in a compressor.

Nominal conditions

Analysis of the GRAZ cycle under nominal conditions was conducted for the following values of parameters:

- Compressor stages group internal efficiency: 90%
- Turbine stages group internal efficiency: 90%
- Combustor efficiency: 99%
- Heat exchanger pressure loss: 4.3%
- Combustor pressure loss: 5%
- Pump efficiency: 90%
- Electric generator efficiency: 99%
- Cycle overall mechanical efficiency: 99%

- Overall power output: 500 MW$_e$
- Temperature after combustor: 1,700°C
- Condenser pressure: 0.005 MPa
- Condensate temperature: 33°C.

Working medium parameters were calculated based on the NIST/ASME steam property tables.

The GRAZ cycle performance under nominal conditions was analysed basing on maximal overall thermal efficiency criteria (minimum 60% HHV for the 500 MW$_e$ class unit).

It was possible to increase the GRAZ cycle efficiency by adding a compressor inter-stage cooling and recuperation (high-temperature regeneration) to the basic configuration. Nevertheless, it should be pointed out that implementing the GRAZ cycle would require introducing, in a different range, cooling systems of the hottest elements, including turbine blades. Usually implementation of such cooling radically decreases overall cycle efficiency, so it is especially important that the cycle has certain efficiency "reserves" (above 60% HHV).

During this research, a modified version of the GRAZ cycle was chosen. The GRAZ cycle was additionally equipped with the following items:

- An inter-stage cooling system by condensate injection in the steam compressor
- Additional heat exchanger – regenerator – to heat steam flow at the inlet to the combustor
- "Classic" two-stage regenerative heating system with a deaerator.

In order to analyse cycle performance under changed conditions, a simple version of the system was chosen (Fig. 18). This way, we can concentrate on basic issues. Operation of cooling system and additional regeneration system under changed conditions can be taken into account in the next step.

Off-Design Analysis

Part-load analysis was an important issue and it should be taken into account when designing and defining operational characteristics. Results of the cycle behaviour analysis under part-load conditions should support defining the cycle structure and its nominal parameters, as well as design solutions and characteristics of a given subsystem. It could happen that calculation results based on nominal conditions analysis (e.g. by use of maximum efficiency criteria) are not useful from the operating point of view. Some specific working conditions of the cycle performance could be assumed, making proper operation possible only in a very narrow range of parameters, different from nominal ones, making the cycle not adaptable to power output changes. Thus, in the extreme case, starting up the cycle would practically not possible at all. It should be stressed that no complex analysis of the GRAZ cycle part-load operation has been published so far.

Part-load operation characteristic research can be reduced mainly to conditions of co-operation between turbomachinery, the heat exchange element and other equipment. The regenerative heat exchangers included in the GRAZ cycle provide a kind of link between low- and high-pressure part of the cycle. A specific feature of this study was the existence of many bonds and limits. Bonds were defined mainly by the cycle configuration and properties of devices, together with their characteristics. Limits usually result from boundary values of operating parameters. Thus, studying conditions of co-operation of the cycle can be reduced to description and analysis of all possible operational conditions for which bonds and limits were fulfilled.

Basic off-design analysis of the GRAZ cycle developed here was limited only to a single-shaft version with constant rotational speed. So it was necessary to stress that a condensing turbine module (LPT) was formally isolated, though according to the concept of dividing the GRAZ cycle into low-pressure and high-pressure part, the LPT module was assigned to the low-pressure part (LP). Isolating the LPT module was done for two reasons:

1. The analysed GRAZ cycle concept was a single-shaft variant, so the LPT turbine was placed in sequence power generation in mechanical sense, and was one of elements when summing up mechanical power generated on shaft-ends of particular turbine stage group.

2. Because of using the same elementary mathematical model (turbine stage group), describing the module's performance.

Mathematical model of the GRAZ cycle was formulated as a very complex and strongly non-linear algebraic equation set. Due to the complex structure of the GRAZ cycle, solving this model was difficult. There were interconnections between main flow streams (collectors, distributors in the flow stream paths), while a split ratio was unknown and needed to be found.

The equation set should be divided in such a way that equation sub-sets would be connected by possibly minimal number of common variables – so called coordination variables. Additionally, equation sub-sets should be coherent with mathematical models of particular devices. In this way it is possible to formulate general modules modelling devices' performance which makes a possibility to study different structures made of typical modules.

Mathematical models of individual elements of the system (modules) were presented by Kiryk (2002) and Miller et al. (2001). In particular, a model of turbine stage group was proposed by Miller et al. (2000).

Mathematical model of a steam compressor requires some additional remarks. Such a compressor is a new and virtually unknown element introduced in the GRAZ cycle. To our knowledge, this type of compressor has not been used until now, and its implementation would require separate design work. For the purpose of mathematical modelling, several assumptions were made, on the basis of the analogy of flow properties of superheated steam and gas, taking into consideration the possibility of applying several design ideas known from air compressors. Fig. 19 shows compressor characteristics used in the research.

In case of a compressor operating with constant, rated rotational speed – and simulating GRAZ cycle operation – the compressor characteristic is needed only to find out, whether operating conditions being examined (reduced flow and compression) are attainable by changing pitch of blades and vanes and how it will reduce the compressor efficiency.

It seems that under the circumstances assuming compressor characteristic of Fig. 19 is acceptable.

Performance static characterisitcs

In order to solve the mathematical model of the GRAZ cycle, a specific sequence-iteration program was developed based on modelling experience for "classic" condensing turbine sets.

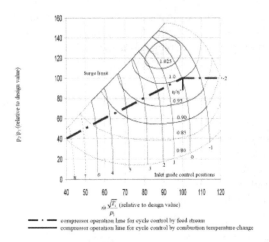

Fig. 19. Compressor characteristic: variable inlet vane angle

The GRAZ cycle power output was changed affecting main circulating mass flow (feed steam flow change) or changing temperatures after combustor (combustion temperature change), or both (combined method). The first case could be treated as a quantitative control method, the second one as a qualitative one and the third as a combined quantitative and qualitative.

Control was possible on the working medium mass flow side (by compressor outlet mass flow) and by changing flow of hydrogen-oxygen mixture supplied to combustors (fuel mass flow). When changing the main circulating mass flow, pressures change accordingly. To follow these pressure changes, the cycle should be equipped with two valves – at the inlets of HHPT and LPT turbines, respectively. There is a kind of analogy here with the accumulation operating mode. The influence of this control concept on the GRAZ cycle performance was simulated using the programme under different part-load and overload conditions. Calculations, if possible, were performed within the range 20–120% of the nominal point.

When performing calculations, the following constraints and limits were taken into consideration:

1. Compressor characteristics working range for the variable inlet vane angle.

2. Maximal temperature of the cycle (combustor outlet temperature 1973 K).

3. Minimal temperature difference in the heat exchanger, to ensure proper heat exchange conditions.

4. Stoichiometric ratio of fuel and oxygen, to ensure stability of the combustion process.

Input data for these calculations were:

1. Compressor outlet mass flow.

2. Combustor outlet temperature.

3. Condenser cooling conditions.

Apart from sequential calculations, iteration loops were required to define operating conditions of heat exchangers (EX1, EX2) and operating conditions between the compressor (COMP) and the high-high-pressure turbine (HHPT).

Combustion temperature change characteristics show that it was possible to operate only within very limited range of changes ($P/P_0 \subset (0.65, 1.0)$). Change of combustion temperature with constant feed steam flow causes almost-linear power output change.

Feed steam flow change with constant combustion temperature was accompanied by virtually linear change of power output, as well as linear pressure change in the cycle. In this case, control was possible down to 37% of P/P_0 and in overload conditions. Overall thermal efficiency varies only slightly within a wide range of the power output values (it was almost constant). Stable overall thermal efficiency is a very important characteristic of the GRAZ cycle. The overall thermal efficiency change is higher when control is exercised by feed steam flow change.

Combustion temperature change has a smaller impact on changes of the overall thermal efficiency. However, it could be applied within very limited operational range of the cycle performance. Hence a combined method could be considered – the main control would be done by changing feed steam flow together with changing combustion temperature for compensation regulation.

Cooling system concept

Possible solutions of turbine cooling systems for the GRAZ cycle are intensively researched by Hitachi, Mitsubishi, Toshiba and other R&D centres Desideri et al. (2001). According to Hitachi Kizuka et al. (1999) the efficiency of the cycle without cooling system is 61.3% HHV, for the cycle with closed water-steam cooling system is 60.1% HHV, for the cycle with closed steam cooling system is 58.7% HHV and for the cycle with open steam cooling system is 54.7% HHV. Cooling system implementation in those cases has decreased the HHV efficiency respectively about 1.2, 2.6 and 6.4 percentage point. Negative impact of cooling implementation is especially big for the open cooling system.

The Mitsubishi Co. has assessed the efficiency of the cycle with the cooling system as 61.8% HHV Sagae et al. (1998) which was then corrected to 60.8% HHV Mouri (1999) and the Toshiba Co. has set this value to 60.1% HHV Okamura et al. (1999). Estimates, which have been recently presented, indicate a slightly higher numbers 61.1–61.8% HHV. Some technology solutions of turbine blades' cooling and results of their analysis were published in Mouri & Taniguchi (1998); Okamura et al. (1999); Sagae et al. (1998).

Summary

Specific calculation programme has been developed for the GRAZ cycle to define static characteristics and part-load analysis under different off-design working conditions. This analysis was done for the single-shaft version with constant rotational speed. It was found that GRAZ cycle has good operational properties for part-load, stable efficiency and acceptable properties for overload. Results obtained for the GRAZ cycle basic off-design performance characteristics seem to be, in our opinion, the first results published in this field.

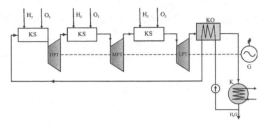

Fig. 20. The MNRC cycle diagram.

3.2.5 Modified New Rankine Cycle – MNRC

The MNRC is proposed by the authors after reconciliation of the abovementioned solutions. The main features increasing the cycle's efficiency are single energy input points (exchanger and one H_2-O_2 combustor) and two-stage re-heat (two H_2-O_2 combustors).

The MNRC cycle is a new concept developed as a result of analysis of previously developed solutions. Contrary to other cycles, in the MNRC solution the heat recovery steam generator (HRSG) is located downstream from the last turbine stage group, within the low-pressure part. This concept has been already mentioned in elaboration Bannister et al. (1997). Arrangement of the cycle allows adding another reheat stage before the low-pressure turbine stage group. As a result a very high thermal efficiency of the cycle is achieved. Flow diagram of the MNRC cycle is shown in Fig. 20. Pressure split between high-pressure and mid-pressure turbine stage groups was optimised according to maximum overall HHV thermal efficiency. Working medium parameters were calculated using the NIST/ASME steam property tables.

The MNRC cycle is relatively insensitive to variation of main parameters, in particular it allows to reduce combustion temperature from 1,700°C to 1,300°C, while maintaining the required 60% HHV overall thermal efficiency.

Static characteristics

Part-load analysis is an important issue for any type of cycle including the HFCTC and it should be taken into account during designing and defining operating characteristics. Results of the cycle's behaviour analysis in part-load conditions should support defining of the cycle's structure and its nominal parameters as well as design solutions and characteristics of components. It could happen that calculation results based on nominal conditions analysis (e.g. by use of maximum efficiency criteria) are not useful from the operating point of view. Some specific working conditions of the cycle performance could be assumed, making proper operation possible only in a very narrow range of parameters, different from nominal ones, making the cycle not adaptable to power output changes. Thus, in the extreme case, starting up the cycle would practically not possible at all. It is to be stressed that no complex analysis on the HFCTC's part-load operation has been published so far.

Part-load operation characteristic research of the HFCTC can be reduced mainly to study conditions of co-operation among rotary machines and the heat exchange element (it means the HRSG or regenerative heat exchangers set), which is a link between low- and high-pressure parts of the cycle, and other equipment. A characteristic feature of this study is existence of many bonds and limits. Mainly the cycle's configuration and properties of devices together with their characteristics define bonds. Limits arise usually from boundary values of working parameters. Study on conditions of co-operation of the HFCTC can be reduced then to description and analysis of all possible operational stages in which bonds and limits are fulfilled.

Mathematical models of the HFCTC are formulated as very complex and strongly non-linear equation set. In order to solve effectively the MNRC cycle's model a specific sequence-iteration program was developed based on modelling experience for "classic" condensing turbosets.

Control over main circulating mass flow or adjustments of temperatures after combustors (combustion temperatures) or both (a combined method) may be used to change the MNRC cycle's power output. The first case could be treated as a quantitative control method, the second one as a qualitative and the third as a combined quantitative and qualitative. These control methods could be practically be implemented without any valves operating within the main circulating mass flow because of its high temperature. Control is possible on the feed water side (by feed water pump) and by change of hydrogen-oxygen mixture supplied to combustors. When changing the main circulating mass flow, pressures change accordingly. There is a kind of an analogy here to accumulation work mode. An influence of this control concept onto the MNRC cycle's performance was simulated using the programme in different part-load and over-load conditions.

As it is shown, a change of the main circulating mass flow at constant (nominal) combustion temperatures, is accompanied with a practically linear change of an power output (both overall and on individual turbine stage groups) as well as linear pressure change in the cycle. Internal efficiencies of the HPT and MPT are constant while the internal efficiency of the LPT varies. In this area the HFCTC displays behaviour similar to the "classic" steam cycle. However the cycle's overall thermal efficiency behaves differently. It varies only slightly in a wide range of the power outputs (it is almost constant). A stable overall thermal efficiency is very important characteristic of the HFCTC.

Changes of combustion temperatures, with the main circulating mass flow set constant, also result with almost-linear changes of the power output (both overall and on individual turbine stage groups). The overall thermal efficiency change is higher than in case of flow-control mode. A combined method is possible to consider then – the main control by the main circulating mass flow with additional combustion temperature control for larger power output increments.

Cooling system concept

The high-pressure turbine should be primarily analysed when considering engineering feasibility of the MNRC cycle. The middle-pressure turbine's working conditions are close (slightly lower) to the high-pressure turbine in the GRAZ cycle that had been already analysed in the WE-NET Program working-task frame.

An identification of possible cooling systems for the MNRC cycle, and their influence on the efficiency, requires defining dimensions – specifically those of turbines – in order to evaluate a surface that would need to be cooled down. An example of blading design for the high-pressure turbine (HPT on Fig. 20) is proposed. While this solution should not be considered final or even optimal, it still could be a good basis for assessments of HPT turbine design possibilities. The HPT turbine should be a high-speed machine (7,000–12,000 rpm), with 5–10 stages, installed in one-, two- (most probably) or three-casing mount. It seems that it should be a very compact machine of blading's diameter below 1.0 m and length 1.5–2.0 cm.

It is necessary to underline that high inlet parameters of the working medium allow to significantly increase the turbine stages' load – isentropic enthalpy drop of 160–230 kJ/kg (compared to 40–80 kJ/kg for conventional turbine stages) – while keeping Mach numbers at reasonable levels of 0.4–0.5. Such conditions would result with power of 12–17 MW per stage. An interesting feature is that as opposed to traditional turbines, there is no radical temperature drop of the working medium along the HPT turbine flow part, so cooling is required on the whole length of the HPT turbine. Blade channel divergence is small (inlet/outlet blades' height ratio is 3.5–4).

A concept of the cooling system for the MNRC-HPT turbine was based on results obtained for the GRAZ cycle by Hitachi Ltd. Kizuka et al. (1999); Mouri (1999). Results were transformed to higher parameters and smaller dimensions of the MNRC-HPT turbine flow part.

Following assumptions were made when evaluating required output of the cooling system for the first turbine stage:

- The highest accepted temperature of blade metal is $1,000°C$,

- Temperature drop on TBC (thermal barrier coating) is $400°C$.

an average heat flow on TBC's border is $2 \, MW/m^2$, initial temperature of medium used for cooling is $250–300°C$. Next stages require adequately lower cooling capacity. Total demand for cooling medium is about 12% of the working flow. Taking into account results achieved by Hitachi, Mitsubishi and Toshiba Kizuka et al. (1999); Mouri (1999); Okamura et al. (1999); Sugishita et al. (1996); Uematsu et al. (1998) it seems realistic to achieve similar parameters in a short-term run. Mass and energy balance of the MNRC cycle with cooling system was defined according to the abovementioned evaluation. It was assumed that, similar to other concepts, cooling is carried out with steam obtained from condensate in the regenerative heat exchanger (Heat Regenerative Steam Generator – HRSG). Cooling steam streams have different pressure levels adequate to individual demands of particular turbine stage groups. As it turns out, there are certain limits of balanced cooling steam amounts, it means 12% – HPT, 13% – IPT and 15% – LPT. When larger flows are needed, water injection into high-pressure combustor is necessary, additionally affecting overall cycle efficiency. A closed counter-current cooling system was not investigated here because of difficulties in obtaining required hermetic conditions of the HPT turbine's cooling system and resulting big pressure differences. Therefore, depending on a cooling system variant, its presence has an important impact on the cycle's efficiency drop. For example:

- Cooling with $300°C$ steam and steam supply to the high-pressure combustor with the following cooling steam ratios HPT – 12%, IPT – 13%, LPT – 15% (those are the highest amounts of cooling steam that are possible to generate) and the cycle's efficiency is 60.9% HHV ($\Delta = 5.5$ percentage points)

- Cooling with $300°C$ steam and water injection ($70°C$) into the high-pressure combustor with the following cooling steam ratios HPT – 13%, IPT – 20%, LPT – 26% and the cycle's efficiency is 57.9% HHV ($\Delta = 8.5$ percentage points),

- Cooling with $300°C$ steam and with the following cooling steam ratios HPT – 13%, IPT – 20%, the low-pressure combustor is switched off and the cycle's efficiency is 60.4% HHV (above 60%HHV requirement),

- Cooling with 300°C steam with the following cooling steam ratios HPT – 13%, cooling stream IPT – 20% is reverted in counter-flow to the intermediate-pressure combustor, the low-pressure combustor is switched off and the cycle's efficiency is 61.9% HHV which is a very promising result showing a direction of the future research.

Technology and design possibilities of MNRC cycle implementation (concerning turbines design) seem to be fully realistic, especially bearing in mind achievements for the GRAZ cycle, and at the same time enable achieving very high efficiency of the cycle that ensures its competitiveness comparing to other concepts.

3.2.6 Comparison of HFCT cycles in nominal conditions

Comparison and cycles' evaluation could be done from different points of view. Basic criterion here is maximal overall thermal efficiency, at least 60% HHV for a 500 MW class unit. Main performance parameters of analysed cycles are shown in Table 2.

Parameter	Graz	Toshiba	Westinghouse	MNRC
p_{max}, bar	350	380	250	250
t_{max}, °C	1,700	1,700	1,700 / 1,600	1,700
Inner Power, MW	513	513	513	513
η_{LHV}, %	70.8	71.2	74.0 / 72.8	79.0
η_{HHV}, %	59.5	59.8	62.2 / 61.2	66.4
Specific Power, kJ/kg	2,202	3,331	3,489	4,706
Electrical Power, MW	500	500	500	500
$\eta_{el,LHV}$, %	69.0	69.4	72.2 / 71.0	77.0
$\eta_{el,HHV}$, %	58.0	58.3	60.6 / 59.7	64.7
Temperature at the most thermally loaded element, °C	1,700	1,700	1,700 / 1,600	1,700
Pressure at the most thermally loaded element, bar	50	73	250	250
Pressure at the most pressure loaded element, bar	350	343	277	277
Temperature at the most pressure loaded element, °C	650	876	517	463
Heat exchanged / HRSG heat load, MW	315	329	256	165

Table 2. Main performance parameters of the HFCT Cycles

As it is shown, efficiencies of GRAZ and TOSHIBA cycles are practically on the limit of the WE-NET Program requirements. The WESTINGHOUSE cycle fulfils these requirements with some margin and the MNRC cycle is far above needed values. It is possible to increase the GRAZ cycle's efficiency by adding compressor inter-stage cooling and recuperation (high-temperature regeneration) to the basic configuration.

Nevertheless, it should to be pointed out that implementation of the HFCT Cycles would require introducing cooling systems of the hottest elements, including turbine blades cooling, operating in a new range of parameters. Usually implementation of such cooling radically decreases overall cycle's efficiency so it is especially important that the proposed concept of the HFCTC has a certain "reserves" of its efficiency (above 60% HHV). It should be said that the MNRCycle has the highest chances to be practically realised due to achieved efficiency. A modified version of the GRAZ cycle was chosen as a leading concept for the Japanese WE-NET programme. The GRAZ cycle was additionally equipped with the following items:

- An inter-stage cooling system by condensate injection in the steam compressor
- Additional heat exchanger – regenerator – for heating of a steam flow at the inlet to the combustor
- "Classic" two-stage regenerative heating system of combustor outlet flow with a deareator.

These versions of the GRAZ cycle were named: Inter-cooled Topping Extraction Cycle and Inter-cooled Topping Recuperation Cycle. Evaluation of efficiencies of these GRAZ cycle's versions (without low-pressure regeneration) has resulted with HHV efficiencies 60.13% and 61.45% respectively Desideri et al. (2001); Uematsu et al. (1998).

All cycles discussed achieve very high specific power (related to a maximal mass flow in the cycle). Those values are much higher (even by an order of magnitude) than ones achieved by contemporary heavy-duty gas turbines (450–500 kJ/kg), GTCC units (600–700 kJ/kg) or steam turbo-sets (1,200–1,400 kJ/kg). It would be possible then to build extremely compact power units with minimal usage of structural materials.

When evaluating technical feasibility of analysed cycles the first step should concern nodes (elements) working at the highest levels of steam parameters. A bottom part of the Table 2 refers just to the most loaded elements' analysis. As it is shown, for the WESTINGHOUSE and MNRC cycles (the most efficient cycles) the highest working medium's temperatures occur together with the highest pressures. This refers to high-pressure combustors and high-pressure turbines and these elements seem to be the "critical" ones. Bearing that in mind, the high-pressure combustor was omitted in the TOSHIBA cycle what resulted, however, with setting very high pressure of the cycle, adding high-temperature heat exchangers and lowering the cycle's efficiency. An advantage of the GRAZ cycle, without any doubts, is relatively low pressure in the highest temperature range – as in a "classic" gas turbine unit. On the other hand big disadvantages are:

- The steam compressor, which is a completely unknown element so far
- Very complex system of heat exchangers and
- Generally high complexity of the cycle.

Summarising the above and considering performance in nominal conditions, the most interesting proposals are the GRAZ cycle and the MNRC.

Cycle	η_{HHV}, %	Deviation from reference value ($\eta_{HHV} = 60\%$), %
Graz cycle	59.466	-0.534
Toshiba cycle	59.780	-0.220
Westinghouse cycle	61.125	1.125
Modified New Rankine cycle	65.098	5.098

Table 3. Comparison of HHV-efficiencies for different cycles' configuration

3.3 Summary

All the HFCTC cycles with direct firing which were studied within this research project (500 MW class units) have shown very high thermal efficiency (minimum 60% HHV, 71% LHV) which is far above the performance of the currently used cycles.The possibility to achieve the efficiency at least 10 percentage points higher than the efficiency of the most efficient contemporary power units (which is about 20% more) has been fully confirmed. Similarly all the HFCTC cycles have very high specific power (2,200–4,700 kJ/kg), which is many times higher (up to an order of magnitude) than performance of the contemporary gas or steam turbines or combined cycles. Thus the new concepts could allow to construct extremely compact power units with minimal usage of construction materials. Considering, additionally, a fact that the HFCTC cycles almost totally eliminate CO_2 and NO_x emissions, this solution can be recognised as an interesting alternative for the future power technology development trend when compared to conventional power technologies.

The concept favoured by WE-NET Program at present – the GRAZ cycle – has certain advantages. One of them is relatively low pressure level in high temperature zones: 5 MPa at 1,700°C. However, the GRAZ cycle has also serious disadvantages, like usage of a steam compressor which is a completely unknown element so far, very complex system of heat exchangers and generally high complexity of the cycle.

The MNRC cycle – proposed in the this research – could be attractive in this context, especially because its efficiency is much higher (66% HHV, 77% LHV) and its general structure is simple. A comparative study has shown that it seems fully realistic to implement an effective cooling system for this case.

A sensitivity study undertaken for the MNRC cycle has shown its big "resistance" to any changes of operating parameters. For example, it is possible to reduce working temperature even to 1,300°C while adhering to 60% HHV efficiency requirement.

Specific calculation programmes have been developed for both the GRAZ cycle and the MNRC cycle to define static characteristics and carry out part-load operation analysis in different off-design working conditions. The GRAZ cycle's operating features have proved to be "acceptable", while the MNRC cycle has displayed high operational and control flexibility combined with very stable thermal efficiency at the same time. Taking the above into account, it seems that the MNRC cycle should be more thoroughly investigated in the future.

It is to be stressed that, in opinion of authors, obtained results of the HFCTC's off-design performance characteristics seem to be the first results published in this field.

Further research should focus on general modification of the HFCTC in order to utilise natural gas as a fuel with CO_2 separation. This research direction seems to be natural due to some delay in research on industrial-scale hydrogen production technologies (including hydrogen-fuelled power sector needs). In more detailed scope – it would be correct to continue research on the GRAZ cycle as well, in order to improve its operational flexibility (multi-shaft cycle concept, etc.).

Hydrogen combustion power systems could be an important alternative to the conventional fossil fuel combustion power systems in the future. Hydrogen could be available in large quantities and its utilisation in power generation systems will have minimal influence on the environment.

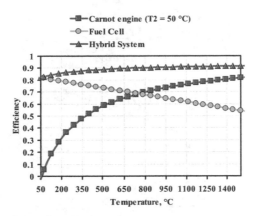

Fig. 21. Theoretical efficiency of a fuel cell – gas turbine system Milewski et al. (2011)

However development of the hydrogen utilisation technologies for energy generation requires R&D work to be undertaken in many fields. Results of the research completed so far show that power generation systems with hydrogen combustion can become competitive in the power technologies market when compared to today's state-of-the-art conventional systems. Nevertheless, high costs of hydrogen production cause make the cycle overall thermal efficiency over 60% (HHV) an essential requirement for this technology.

4. Fuel cell – gas turbine hybrid systems

The main advantage of combining a fuel cell with a gas turbine is that one can create a binary system which can potentially achieve ultra-high efficiencies (see Fig. 21). This task is fulfilled through the other system using the fuel cell exhaust heat.

A typical Fuel Cell-Gas Turbine Hybrid System (FC-GT) consists of the following elements:

- Air Compressor
- Fuel Compressor
- Gas Turbine
- Air Heater
- Fuel Heater
- Fuel Cell.

The Fuel Cell is not the only power source in the SOFC-GT hybrid system (additional power is produced by the gas turbine unit). FC-GT hybrid system efficiency is defined by the following relation:

$$\eta_{HS} = \frac{P_{FC} + P_T - P_C - P_{fuel}}{\dot{m}_{fuel} \cdot LHV} \tag{11}$$

FC-GT hybrid systems can be classified according to their fuel cell module operating pressure. Systems in which the pressure of leaving the fuel cell is comparable to atmospheric pressure

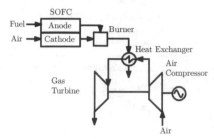

Fig. 22. Diagram of an atmospheric SOFC with a bottom cycle based on an air turbine.

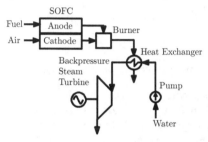

Fig. 23. Diagram of an atmospheric SOFC with bottoming cycle based on an open steam cycle

are called atmospheric FC systems. The second group consists of systems in which the pressure of the exhaust gas leaving the fuel cell is significantly higher than atmospheric; those kinds of systems are called systems with pressurised FC.

This classification determines the location of the fuel cell in a power system. Generally, a fuel cell can perform function similar to a classic combustion chamber, i.e. oxidise fuel supplied to the system, which results in relatively large quantities of electricity being taken from the fuel cell itself. A combustion chamber works with lower amounts of fuel and does not require a large excess of air in order to reduce the temperature of gas getting to the turbine.

4.1 SOFC-GT hybrid systems

4.1.1 Atmospheric SOFC-HS

In case of an atmospheric SOFC, increased system efficiency can be achieved only by heat recuperation through a bottoming cycle. In systems containing an atmospheric SOFC, additional equipment is needed to recover part of the exhaust heat. The simplest bottom cycle is based on an air turbine subsystem (composed of an air compressor and air turbine). This solution is shown in Fig. 23. Heat is transferred to the bottoming cycle by adequate heat recuperative heat exchangers. In this case, SOFC flue gas is considered as the upper heat source for the air turbine cycle.

Regenerative heat exchangers can be used both on the air and fuel flows. One limitation in this solution is the air compressor, because the temperature of compressed air is increased. Therefore, the desirability of a regenerative heat exchanger at the flow depends on the temperature of the exhaust leaving the gas turbine. This problem can be solved by introducing an additional combustion chamber, located upstream from the exchangers. This makes the fuel cell operation independent from temperature rise before the heat exchangers.

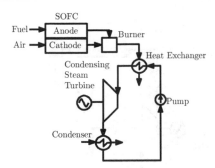

Fig. 24. Diagram of an atmospheric SOFC with bottoming cycle based on a closed steam cycle.

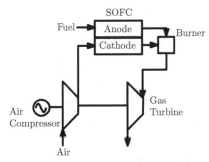

Fig. 25. SOFC-GT hybrid system with the fuel cell placed instead of the combustion chamber.

This system suffers from relatively low efficiency due to the relatively large amount of power consumed by the air compressor. To increase the efficiency of the system, water (steam) can be substituted for compressed air as a working agent. Much less energy is required to increase water pressure than to compress air to the same pressure. Such a system system is illustrated in Fig. 23. The system with a bottoming cycle based on an open steam turbine cycle has better performances than the one based on an air turbine due to the reduced compression work and better heat transfer in the heat recovery steam generator (HRSG).

Steam at a temperature of more than 100°C is discharged from the system. Firstly, the system in this configuration will consume huge amounts of water. Secondly, all the water evaporation heat is lost to the environment. Part of this heat can be recovered using a system working in the Rankine cycle. This solution is presented in Fig. 24 and is comparable to the classic gas turbine combined cycle (GTCC) in which the SOFC is placed instead of a gas turbine.

4.1.2 pressurised SOFC-HS

From the gas turbine system point of view, the pressurised SOFC can be substituted for the combustion chamber.

Simple addition of a gas turbine subsystem to the SOFC raises efficiency to 57%. The power generated by the gas turbine subsystem represents about 30% of the total system power.

The upgrade of the SOFC-based system is the addition of a heat exchanger, which is placed between the fuel cell stack and the air compressor, similarly to upgrade of an open cycle gas

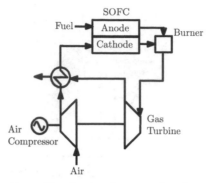

Fig. 26. SOFC-GT hybrid system with a heat exchanger placed upstream from the fuel cell.

Parameter	GT	SOFC	SOFC-GT
Fuel	CH_4	H_2	CH_4
Efficiency, %	20 (27*)	37	66
TIT, °C	1,100	–	1,100
Fuel cell temperature, °C	–	800	800
GT pressure ratio	11	–	9.6
Power given by GT in relation to total system power, %	100	0	22

Table 4. Main parameters of a SOFC-GT Hybrid System.
*with heat recuperation

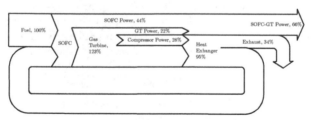

Fig. 27. Sankey's diagram of energy flows of SOFC-GT system.

turbine system (see Fig. ??). The heat exchanger is fed by the gas turbine outlet stream. An adequate heat exchanger may increase the efficiency by recovering part of the gas turbine exhaust heat. This solution increases efficiency to 66% and decreases gas turbine pressure ratio to 8.2. TIT is still relatively high at 1,100°C.

The addition of a SOFC to a gas turbine system can boost efficiency to 57%. By changing the fuel from hydrogen to methane efficiency is increased to 63% due to reforming reactions which convert thermal energy into the chemical energy of fuel. The addition of a heat exchanger increases the efficiency even further, to 66%.

Energy flow paths in a pressurised SOFC-GT hybrid system are shown in a form of Sankey diagram in Fig. 27.

Selection of the size, design and working parameters of the SOFC is crucial when seeking to obtain a highly-efficient hybrid system. Additionally, both the gas turbine and the air compressor should be designed for operation with SOFC.

4.1.3 Control issues

It should be underlined that in case of a system which contains both a pressurised SOFC and a gas turbine, varying the amount of fuel injected is not the only way to control the power output. Fuel cell voltage and current are quite dependent on the variable rotational speed of the compressor-turbine unit. This is accompanied by varying system efficiencies. Hence there is a need to formulate an appropriate control concept (control strategy logic) and its approach for technical realisation.

Control strategy is an important element in designing any system of this kind and it constituted a significant part of the modelling works done. Off-design (part-load) analysis is an important issue for any type of system involving an SOFC-GT hybrid solution and should be taken into account when designing and defining the operational characteristics. A proper off-design performance map underscores control strategy design. Results drawn from system behaviour analysis under part-load conditions should aid in defining the system structure and its nominal parameters, as well as the design solution and characteristics of a given subsystem.

Part-load operation characteristics research regarding SOFC-HS can be reduced mainly to study of the conditions of co-operation among the SOFC, turbomachinery and other equipment. A specific feature of this study is the existence of many bonds and limits. Bonds are defined mainly by the system configuration and properties of devices that make up the system, together with their characteristics. Limits usually result from boundary values of working parameters. Thus, studying the conditions of co-operation of the SOFC-GT can be reduced to describing and analysing all possible operational conditions. Part-load and overload performance characteristics of SOFC-GT were calculated and analysed to show control possibilities of the cycle.

Off-design (part-load) analysis is an important issue for any type of system including an SOFC-GT and should be taken into account during designing and defining operational characteristic. Results of the system behaviour analysis under part-load conditions should aid in defining the system structure and its nominal parameters, as well as the design solution and characteristics of a given subsystem.

Implementation of a control system depends on its structure, which is subject to regulatory processes at work in varied conditions. Two sources of electricity may exist in hybrid systems. Usually one of these sources is to provide the main energy flux while the other plays auxiliary role. There is therefore a need to identify possible structures for systems analysis of power distribution to the various subsystems. A multi-stage procedure is used to determine the control strategy of SOFC-GT hybrid systems, as follows:

1. Internal constraints of the system are determined based on the chosen structure and parameters of the nominal SOFC-GT systems (i.e. mechanical, flow and electrical connections between the elements and characteristics of specific elements).
2. Mathematical model of the system is built at the level of off-design operation.
3. Based on the model all possible operating conditions are determined. A database of those steady-state points of the system is created.

4. Based on the database an adequate searching algorithm is used to find the most convenient operation line of control strategy for chosen criteria (e.g. highest system efficiency).

5. The control strategy is realised by the first level of a multi-layer control system.

The control strategy of the system should allow to quickly and accurately follow the load profile, while maintaining good system efficiency. Safe operation of the system is an obvious requirement. Based on mathematical modelling and numerical simulations, the control strategy for an SOFC-GT is presented.

An SOFC-GT hybrid system has three degrees of freedom, and with regard to m-fan up to four; which means that at least three free parameters can vary independently within certain ranges. Any combination of them should define a certain state of the system. An SOFC-GT hybrid system has only one control variable, which is generated power, and three manipulated variables, i.e. the current cell module, the flow of fuel and the electric generator load. The dependencies and relationships occurring in the hybrid system can be divided into four groups:

1. Correlations defined by a mechanical scheme of the system (mechanical linkage of turbines, compressors, generators).

2. Correlations defined by the flow chart, (i.e. flow relationship between the compressor, turbine, fuel cell, heat exchangers and other components of the system and the order of the working flow direction through system elements).

3. Depending on specific characteristics of turbines and compressors (average parameters at the inlet and outlet of the rotating machine are closely linked through its characteristics).

4. Depending on the characteristics associated with other system elements.

5. Depending on specific electrical connections inside the system.

An SOFC-GT hybrid system can be controlled using the following parameters:

- Electric current taken from SOFC stack by external resistance (load)
- Fuel mass flow through a valve
- Rotational speed of the compressor-turbine subsystem by power output of electric generator with adequate power electronic converter.

Ensuring safe operation of the system requires elimination of events (operating conditions) which could damage the system or its components. This represents at the same time a limit on the scope of permissible working conditions (states) of the system. Typical constraints normally are a result of:

- Acceptable working fluid parameters (mainly the highest temperature and pressure)
- Acceptable electrical parameters
- Compressor limits (surge line)
- Critical frequencies of rotating machines
- Acceptable torque values.

Those limitations can be defined arbitrarily, and the values presented below are exemplary only. Based on data calculated for all technically possible conditions, additional limits and bonds were applied on the system efficiency map, and they are as follows:

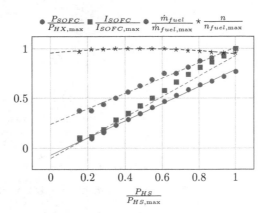

Fig. 28. Relationships between control parameters of a SOFC-GT system. Legend: n – GT shaft rotational speed, I - SOFC stack current, $\dot{m}$ – fuel mass flow, P – overall power output

- Average solid oxide fuel cell temperature $< 1{,}000°\mathrm{C}$
- SOFC stack temperature difference $< 320°\mathrm{C}$
- Turbine Inlet Temperature (TIT) $< 1{,}000°\mathrm{C}$
- Air compressor surge limit curve.

Other conditions reflect the more general requirement to maintain fuel cell stability, fostered by keeping the cell temperature as constant as possible and reducing (limiting) charged current (limited local heat source). The amount of steam at the anode inlet must be monitored to avoid carbon deposition during reforming processes. Deposition takes place when the temperature is too low and/or there is too little steam at the inlet to the stack (low s/c ratio). A reverse flow of gases from the combustion chamber to the anode channels can result in anode coming into contact with oxygen – anode reverse flow can occur with rapid increases in pressure, hence the need to limit the increase in pressure over time. And finally, too low fuel cell voltage can result in unstable fuel cell stack operation.

Taking into account controllable parameters, the control strategy can be based on the following three functional relationships:

$$\dot{m}_{fuel} = f\,(P_{HS})$$
$$n = f\,(P_{HS}) \tag{12}$$
$$I_{SOFC} = f\,(P_{HS})$$

Choosing the highest possible efficiency of the system, the relevant functional dependencies 12 of all controllable parameters take the form of adequate functions of the system power. The controllable parameters of the system as a function of system power are presented in Fig. 28. The figures were normalized to their maximum values. It is evident that both the current cell stack and the amount of fuel supplied are close to the linear trend. The rotational speed of the gas turbine is relatively constant ($+/-$ 10%), reaching a maximum at 50% system load.

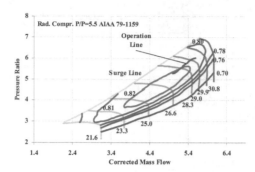

Fig. 29. Operation line of control strategy indicated on air compressor map used during simulations.

Exemplary of the functional relationships (Eq. 12) are as follows:

$$\frac{\dot{m}_{fuel}}{\dot{m}_{fuel,max}} = 0.7725 \cdot \frac{P_{HS}}{P_{HS,max}} + 0.2391$$

$$\frac{n}{n_{max}} = -0.1915 \cdot \left(\frac{P_{HS}}{P_{HS,max}}\right)^2 + 0.1812 \cdot \frac{P_{HS}}{P_{HS,max}} + 0.9561 \tag{13}$$

$$\frac{I_{SOFC}}{I_{SOFC,max}} = 1.0304 \cdot \frac{P_{HS}}{P_{HS,max}} - 0.1001$$

It can be seen that the fuel cell stack current should be increased more rapidly than the fuel mass flow. The GT shaft speed should be increased until system power reaches 50% and, thereafter, the shaft speed has to be decreased. The SOFC-GT hybrid system maintains good efficiency even at partial loads. For example, while reducing the system load to about 40% the efficiency is still over 80% of the nominal value.

Normal operation of the system is possible in a wide power range, from approximately 17% of the rated power. The operation line is located preferably on the characteristics of the compressor – far from the surge limit (see Fig. 29).

Determination of reasonable parameters (due to optimised efficiency) of an SOFC-GT hybrid system also defines the necessary parameters for rotating machinery and other system components. Needed mass flow of both compressor and turbine is in the range 5–9 kg/s at a pressure ratio (both compression and expansion) of 5–6.2. The parameters of rotating machinery needed for the SOFC-GT can be achieved by a single-stage radial compressor and a two-stage axial turbine.

4.2 MCFC-GT hybrid systems

A molten carbonate fuel cell-hybrid system consists of the following elements:

- Air compressor
- Fuel compressor

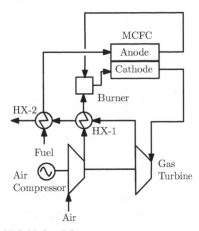

Fig. 30. Pressurised GT-MCFC Hybrid System.

Parameter	Value
Overall Efficiency (LHV), %	58
Re-cycle Factor, %	44
Excess Air Factor	1.68
Average Cell Voltage, mV	659
DC/AC inverter efficiency, %	95
Electric generator efficiency, %	99
Mechanical efficiency of the GT, %	99
Electric motor efficiency, %	95
Turbine Inlet Temperature (TIT), °C	650
Compressor pressure ratio	8.35
Electrolyte material	Li/Na
Matrix thickness, μm	1,000

Table 5. Nominal parameters of a MCFC-HS system.

- Gas turbine
- MCFC module
- Air heater
- Fuel heater
- MCFC module.

Based on mathematical modelling the design point parameters of the system presented in Fig. 30 were estimated. The main system parameters are presented in Table 5.

4.3 Reducing CO_2 emission of gas turbine by implementation of MCFC

The European Union has imposed limits on CO_2 emissions by Member States as a part of its Emission Trading Scheme. This affects fossil fuel power plants to a significant degree as their emissions are governed by the number of emission allowances they receive from the Member State allocation. Excess CO_2 emissions have to be covered by purchasing extra allowances, which is in effect a financial penalty (EUR100/Mg). In contrast, undershooting emission limits enables the operator to sell CO_2 allowances. The selling price of a traded allowance is estimated at EUR20–30/Mg_{CO_2}.

There is a variety of methods available to remove CO_2 from a fossil fuel power plant system Gottlicher & Pruschek (1997). The idea of adopting a molten carbonate fuel cell to reduce CO_2 emissions was developed by Campanari Campanari (2002). An estimation made by Campanari shows that a reduction of 77% in CO_2 emissions can be achieved in a steam turbine power plant.

Fuel cells generate electricity through electrochemical processes Milewski et al. (2011). There are many types of fuel cells, two of them – the Molten Carbonate Fuel Cell (MCFC) and the Solid Oxide Fuel Cell (SOFC) – are high-temperature types. They operate at temperatures ranging from 600–1,000°C. Amorelli et al. Amorelli et al. (2004) described an experimental investigation into the use of molten carbonate fuel cells to capture CO_2 from gas turbine exhaust gases. During experiments performed using a singular cell, an emission reduction of 50% was achieved.

Lusardi et al. Lusardi et al. (2004) investigated the application of a fuel cell system for CO_2 capture from thermal plant exhaust. They found that even without CO_2 separation, the relative emission of carbon dioxide could be reduced below the Kyoto Protocol limit. If a separator is used, emissions could be reduced by 68%.

Use of an MCFC as a carbon dioxide concentrator was investigated by Sugiura et al. Sugiura et al. (2003). In this work the experimental results of CO_2 capture using an MCFC are given. One key conclusion from this work is that the CO_2 removal rate can be obtained by making calculations using electrochemical theory.

Novel methods whereby carbonates were used as an electrochemical pump in carbon dioxide separation from gases were described by Granite et al. Granite & O'Brien (2005).

Based on the review of abovementioned literature, a reduction of at least 50% in CO_2 emissions could be expected.

Hydrogen, natural gas, methanol or biogas may be used as fuels for MCFCs. On the cathode side, a mixture of oxygen and carbon dioxide is required.

An MCFC may work as a carbon dioxide separator/concentrator because the CO_2 is transported from the cathode side to the anode side through molten electrolyte.

The combination of GT unit with MCFC results with a hybrid system (HS) with increased efficiency and decreased carbon dioxide emission. The exhaust flue gas of gas turbine power plant consists mainly of nitrogen, oxygen, steam and carbon dioxide. This mixture can be used as oxidant in the MCFC (cathode feeding). The temperature of the exhaust gas and electric efficiency of GT unit are around 550°C and 35%, respectively. The fuel cells in turn can achieve higher electric efficiency of 50–60%.

Negative ions are transferred through the molten electrolyte. Each ion is composed of one molecule of carbon dioxide, one atom of oxygen and two electrons. This means that an adequate ratio of carbon dioxide to oxygen is 2.75 (mass based) or 2.0 (mole based).

The typical gas turbine flue gas composition is shown in Table 6. The ratio of CO_2 to oxygen is hence 0.25 (mole based) and 0.34 (mass based). This means that flue gas contains an insufficient quantity of oxygen to slow trapping all CO_2.

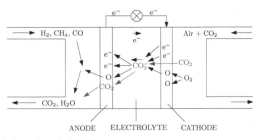

Fig. 31. Working principles of an MCFC.

Name	Value
Air compressor inlet pressure, MPa	0.1
Air compressor inlet temperature, °C	15
Pressure ratio	17.1
Fuel	Natural gas
Fuel mass flow, kg/s	4.0
Turbine inlet temperature, °C	1210
Exhaust gas mass flow, kg/s	213
Turbine outlet temperature, °C	587
GT Power, MW	65
GT Efficiency (LHV), %	33
CO_2 annually emission, Gg/a	250
Relative emission of CO_2, kg/MWh	609
CO_2 mass flow, kg/s	11

Table 6. Nominal parameters of GT Power Plant Granser & Rocca (1996)

The hybrid system (fossil fuel power plant + MCFC) has no obvious objective function of the optimising process, the system performances were estimated based on CO_2 reduction emission factor, which is defined as follows:

$$\eta_{CO_2} = 1 - \frac{m_{CO_2,out}}{m_{CO_2,in}} \tag{14}$$

where: m – mass flow, kg/s; out – MCFC outlet cathode stream; in – MCFC inlet cathode stream.

A GT unit consists of the following elements:

1. Air compressor,

2. Gas turbine

3. Combustion chamber.

Mathematical model of a GT unit was based on the following assumptions:

• Air compressor isentropic efficiency: 79%

• Gas turbine isentropic efficiency: 88%

• No pressure drops across the combustion chamber.

Component	Mass fraction, %	Mole fraction,
CO_2	5.2	3.4
H_2O	4.1	6.6
O_2	15.3	13.6
N_2	74.0	75.4
Ar	1.4	1.0
CO_2/O_2	0.34	0.25

Table 7. Exhaust gas composition.

A commercial gas turbine unit was selected for analysis Granser & Rocca (1996). Nominal parameters of the GT unit and exhaust gas composition are shown in Tables 6 and 7, respectively.

To create a CO_3^{2-} ion, it is needed to split a half mole of O_2 with one mole of CO_2. Adequate mass and molar ratios of CO_2 to O_2 (for capture all carbon dioxide) are 1.38 and 2, respectively. Data given in Table 7 shows that, theoretically, all CO_2 could be captured.

Two cases of gas turbine power plant with the MCFC were investigated. Case 1 concerns a situation when there the GT cycle stays unchanged. It means that MCFC is added at gas turbine's outlet stream. Case 2 concerns the situation when heat exchangers before combustion chamber are added to the gas turbine unit. These heat exchangers are fed by MCFC exhaust streams. Relatively low CO_2 content in flue gas results with low MCFC efficiency. The MCFC efficiency is about 34% (based on Lower Heating Value, LHV). It seems to be unreasonable to combine a low-efficiency MCFC with a high-efficiency Gas Turbine Combined Cycle unit (with efficiency about 55%) so this case was not investigated.

Proper objective function of the optimisation process is not apparent. The MCFC is installed to capture the CO_2, and from this point of view the quantity of captured CO_2 should be maximised. But from the other point of view, the MCFC uses the same fuel as the gas turbine to produce electricity. Both analysed cases were optimised to obtain maximum system efficiency.

In both cases apart from MCFC, following devices should be installed:

1. CO_2 separator (water condensing unit)

2. Catalytic burner

3. DC/AC converter with efficiency of 95%.

The CO_2 separator is cooled by water. When steam condenses, water is separated from the carbon dioxide stream.

The catalytic burner is fed by pure oxygen to utilise the rest of methane, hydrogen and carbon monoxide. An oxygen production (e.g. extraction from air) requires energy input. The production of one kilogram of oxygen at atmospheric pressure requires from 200 to 300 kJ energy input. The value of 250 kJ was taken into calculations, which decreases the system efficiency depending on the amount of consumed oxygen.

Installation of an MCFC at gas turbine outlet means back pressure drop of about 1%. It decreases the efficiency of the gas turbine from 33% to 32%.

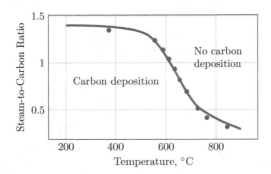

Fig. 32. Minimum temperature and required ratio of steam-to-carbon (s/c) above which no carbon deposition takes place herle et al. (2004)

Name	GT plant	Case 1	Case 2
GT-MCFC power (total power), MW	65	80	77
GT power/total power, %	100	81	82
MCFC power/total power, %	0	19	18
GT-MCFC efficiency (LHV), %	33	33	40
CO_2 emission reduction factor, %	0	73	91
Annual CO_2 emission, Gg/a	250	67	18
Relative CO_2 emission, kg/MWh	609	132	37
MCFC efficiency (LHV), %	–	34	36
GT efficiency (LHV), %	33	32	41
Fuel utilization factor Milewski et al. (2006), %	–	90	90
Average cell voltage, mV	–	513	486
Current density, mA/cm^2	–	29.5	29.6
Oxygen mass flow, kg/s	–	0.2	0.2
MCFC/GT fuel ratio	–	0.52	0.65

Table 8. Nominal parameters of GT-MCFC system.

Case 1

The MCFC is fed by two streams: GT exhaust gas at cathode side and a mixture of methane and steam at anode side.

The MCFC anode outlet stream is directly delivered to the CO_2 separator.

The system was optimised to obtain maximum system efficiency. Primary (adjusted) variables of the optimizing process were:

- Cell current density
- MCFC/GT fuel ratio.

Parameters obtained during the optimizing process are given in Table 8.

A simple combination of the MCFC with GT gives:

- CO_2 emission reduction of 73%
- Unchanged electrical efficiency

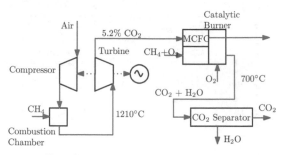

Fig. 33. GT-MCFC system – Case 1

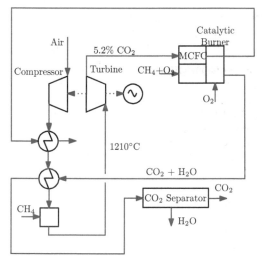

Fig. 34. GT-MCFC system – Case 2

- Power output increase of 23%.

Case 2

The MCFC-GT Case 2 was created by adding two heat exchangers. The heat exchangers are installed to recover exhaust heat from MCFC outlet streams. Note that the GT efficiency would increase with a recuperative heat exchanger when no MCFC is installed as well.

The system was optimised with the same conditions like Case 1. Nominal parameters of Case 2 of GT-MCFC system are given in Table 8.

GT-MCFC Case 2 generates slightly less power in comparison with Case 1. During the simulations a constant value of Turbine Inlet Temperature (TIT) was assumed. The implementation of heat exchangers means lower fuel mass flow demanded by the combustion chamber.

A reduction of the CO_2 emission of 91% is obtained. Simultaneously, electric efficiency is increased to 40% (LHV) what gives the relative emission of CO_2 of 37 kg/MWh.

Summary

The CO_2 emission reduction factor and CO_2 relative emission were used to compare the systems. The MCFC could reduce the CO_2 emission by more than 70% from gas turbine power plant exhaust. The relative CO_2 emission decreases more significant because the MCFC produces additional power.

Relatively low efficiency of the MCFC is caused by low CO_2 content at gas turbine exhaust, which limits maximum cell voltage.

A combination of MCFC with a GT unit means higher investment costs. Other devices like water separator and heat exchangers increase the total investment cost as well. It should be noted that typical CO_2 separation methods also increase the investment costs.

Application of the MCFC in a Gas Turbine plant gives a relatively high reduction in CO_2 emissions. The relative CO_2 emission of the GT unit is estimated at 609 kg_{CO_2}/MWh while in contrast the MCFC-GT hybrid system has an emission rate of 135 kg_{CO_2}/MWh. The quantity of CO_2 emitted by the MCFC-GT is 73% lower than is the case with the GT plant.

Application of the MCFC in a Coal Fired Power Plant gives a relatively high reduction in CO_2 emissions. The relative CO_2 emission of the coal plant is estimated at 1137 kg_{CO_2}/MWh while in contrast the MFCF-CFPP hybrid system has an emission rate of 300 kg_{CO_2}/MWh. The quantity of CO_2 emitted by the MCFC-CFPP is 60% lower than is the case with the CFPP.

As mentioned earlier, all cases were optimised to achieve maximum power generation efficiency. However, this might be changed, if it is assumed that the main task of the MCFC is to limit CO_2 emissions, which would result in the CO_2 emission reduction factor being used as the objective function of the optimisation process. If this factor is optimise,d the cell voltage at last cell can fall below zero and the MCFC will work as a CO_2 concentrator. At the very least, the MCFC would generate no power, and might even consume some. However, the main task of a power plant is power generation; hence hybrid system efficiency was chosen as the objective function for optimisation task.

Important technical issues such as sulphur or dust resistances of the MCFC fell outside the remit of this chapter, although they can evidently limit the application of MCFCs in both coal fired boiler and gas turbine power plants.

MCFCs could be profitably used in existing power plants which have been assigned emission CO_2 limits. MCFCs could potentially decrease CO_2 emissions, leaving the power generation capacity of the system at least the same, if not greater.

5. References

(n.d.). http://www.aist.go.jp/nss/text/outline.htm.

Amorelli, A., Wilkinson, M. B., Bedont, P., Capobianco, P., Marcenaro, B., Parodi, F. & Torazza, A. (2004). An experimental investigation into the use of molten carbonate fuel cells to capture CO_2 from gas turbine exhaust gases, *Energy* 29(9-10): 1279 – 1284. 6th International Conference on Greenhouse Gas Control Technologies.

Baader, G. (n.d.). *GE Aeroderivative Gas Turbines – Design and Operating Features*, reference document ger-3695e edn, General Electric.

Bannister, R. L., Newby, R. A. & Yang, W. C. (1998). Development of a hydrogen-fueled combustion turbine cycle for power generation, *Journal of Engineering for Gas Turbines and Power* 120(2): 276–283.
URL: *http://link.aip.org/link/?GTP/120/276/1*

Bannister, R., Newby, A. & Yang, W.-C. (1997). Development of a hydrogen-fueled combustion turbine cycle for power generation, *ASME International Gas Turbine & Aeroengine Congress & Exhibition*, pp. 97–GT–14.

Campanari, S. (2002). Carbon dioxide separation from high temperature fuel cell power plants, *Journal of Power Sources* 112(1): 273 – 289.

Desideri, V., Ercolani, P. & Yan, J. (2001). Thermodynamic analisis of hydrogen cmbustion turbine cycles, *International Gas Turbine Congerss & Exibition*, pp. 2001–GT–95.

Funatsu, T., Dohzono, Y. & Fukuda, M. (1999). Startup analysis of an H2-O2-fired gas turbine cycle, *Electrical Engineering in Japan* 128(1): 9–16.
URL: *http://dx.doi.org/10.1002/(SICI)1520-6416(19990715)128:1<9::AID-EEJ2>3.0.CO;2-T*

Gottlicher, G. & Pruschek, R. (1997). Comparison of CO_2 removal systems for fossil-fuelled power plant processes, *Energy Conversion and Management* 38(0): S173–S178.

Granite, E. J. & O'Brien, T. (2005). Review of novel methods for carbon dioxide separation from flue and fuel gases, *Fuel Processing Technology* 86(14-15): 1423 – 1434. Carbon Dioxide Capture and Sequestration.

Granser, D. & Rocca, F. (1996). New high-efficiency 70 MW heavy-duty gas turbine, *Proceedings of Power-Gen Conference, New Delhi, India*.

Gretz, J. (1995). Hydrogen energy developments status and plans in europe, *International Hydrogen and Clean Energy Symposium*, pp. 7–13.

Van herle, J., Marechal, F., Leuenberger, S., Membrez, Y., Bucheli, O. & Favrat, D. (2004). Process flow model of solid oxide fuel cell system supplied with sewage biogas, *Journal of Power Sources* 131(1-2): 127 – 141. Selected papers presented at the Eighth Grove Fuel Cell Symposium.

Iki, N., Hama, J., Takahashi, S., Miller, A. & Kiryk, S. (1999). Future hydrogen technologies in power engineering and semi-closed gas turbine system, *Prace Naukowe Politechniki Warszawskiej, z. Mechanika* 181: 97–106.

Jericha, H. (1984). A new combined gas steam cycle promising up to 60% thermal efficiency, *15th International Congress on Combustion Engines*.

Jericha, H. (1991). Towards a solar-hydrogen system, *ASME COGEN-Turbo IGTI* 6: 435–442.

Kaya, Y. (1995). Future energy systems and the role of hydrogen, *International Hydrogen and Clean Energy Symposium*, pp. 3–6.

Kiryk, S. (2002). *Research on Hydrogen Fuelled Gas Turbine Set*, PhD thesis, Warsaw University of Technology.

Kiryk, S. & Miller, A. (2001). Calculation of steam and water thermodynamic properties in high temperature and pressure conditions – ITC-PAR calculation routines, *Scientific Leaflets of the Warsaw University of Technology section Mechanics* 195.

Kizuka, N., Sagae, K., Anzai, S., Marushima, S., Ikeguchi, T. & Kawaike, K. (1999). Conceptual design of the cooling system for 1700°C class hydrogen-fueled combustion gas turbine, *Journal of Engineering for Gas Turbines and Power* 121: 108–115. Transactions of the ASME.

Lusardi, M., Bosio, B. & Arato, E. (2004). An example of innovative application in fuel cell system development: CO_2 segregation using molten carbonate fuel cells, *Journal of*

Power Sources 131(1-2): 351 – 360. Selected papers presented at the Eighth Grove Fuel Cell Symposium.

Milewski, J., Salacinski, J., Miller, A. & Badyda, K. (2006). Influence of the fuel utilization factor on the performance of solid oxide fuel cell hybrid systems, *Chemical and Process Engineering* 27(1): 237–254.

Milewski, J., Swirski, K., Santarelli, M. & Leone, P. (2011). *Advanced Methods of Solid Oxide Fuel Cell Modeling*, 1 edn, Springer-Verlag London Ltd.

Miller, A. (1984). *Gas Turbines and Gas Turbines Combined Cycles [Turbiny gazowe i uklady parowo-gazowe] (in Polish)*, Wydawnictwa Politechniki Warszawskiej.

Miller, A., Kiryk, S., Badyda, K. & Milewski, J. (2001). Research on hydrogen-fuelled gas turbine, *Polish Research Grant KBN 8T10B00918*, Institute of Heat Engineering, Warsaw University of Technology.

Miller, A., Lewandowski, J., Badyda, K., Kiryk, S., Milewski, J., Hama, J. & Iki, N. (2002). New efficient hydrogen-fuelled combustion turbine cycle – a study of configuration and performance, 14th *World Hydrogen Energy Conference*, p. A207a.

Miller, A., Lewandowski, J., Trzcinska, Z. & Abed, K. (2000). Generalized performance characteristics of turbine stage groups. An attempt to supplement the Flugel's-Stodola's law, *Archive of Mechanical Engineering* XLVII(1): 33–52.

Moritsuka, H. & Koda, E. (1999). Hydrogen-oxygen fired integrated turbine systems – comparison on MORITS and GRAZ, *Proceedings of the International Gas Turbine Congress*, number TS-18, pp. 401–404.

Mouri, K. (1999). R&d of 1700°C hydrogen combustion turbine in WE-NET program, *Proceedings of the International Gas Turbine Congress*, number TS-11, Kobe, pp. 351–358.

Mouri, K. & Taniguchi, H. (1998). Research and development of hydrogen combustion turbine and very hot side hear exchanger system in we-net project in Japan, *International Joint Power Generation Conference*, ASME.

Nat (2006). *The Gas Turbine Handbook*.

NEDO (1997). International clean energy network using hydrogen conversion (WE-NET), *Annual summary report on results*, New Energy and Industrial Technology Development Organization (NEDO).

Okamura, T., Kawagishi, H., Koga, A. & Ito, S. (1999). Development of hydrogen-oxygen combustion turbine cooling blades in WE-NET project, *Proceedings of the International Gas Turbine Congress*, number TS-46, pp. 599–606.

Poullikkas, A. (2005). An overview of current and future sustainable gas turbine technologies, *Renewable and Sustainable Energy Reviews* 9(5): 409 – 443.

Sagae, K., Kizuka, N., Kawaike, K., Anzai, S., Marushima, S. & Ikeguchi, T. (1998). R&D for 1700C-class turbine closed loop cooled blades, http://www.enaa.or.jp/WE-NET/ronbun/1998/15/1598.htm.

Sugishita, H., Mori, H. & Uematsu, K. (1996). A study of thermodynamic cycle and system configurations of hydrogen combustion turbines, *Proceedings of the 11th World Hydrogen Energy Conference*, pp. 1851–1860.

Sugiura, K., Takei, K., Tanimoto, K. & Miyazaki, Y. (2003). The carbon dioxide concentrator by using MCFC, *Journal of Power Sources* 118(1-2): 218 – 227. Scientific Advances in Fuel Cell Systems.

Uematsu, K., Mori, H. & Sugishita, H. (1998).   Topping recuperation cycle for hydrogen
    combustion turbine in WE-NET,
    http://www.enaa.or.jp/WE-NET/ronbun/1998/16/1698.htm.

6

Gas Turbine Diagnostics

Igor Loboda
National Polytechnic Institute
Mexico

1. Introduction

A gas turbine engine can be considered as a very complex and expensive mechanical system; furthermore, its failure can cause catastrophic consequences. That is why it is desirable to provide the engine by an effective condition monitoring system. Such an automated system based on measured parameters performs monitoring and diagnosis of the engine without the need of its shutdown and disassembly. In order to improve gas turbine reliability and reduce maintenance costs, many advanced monitoring systems have been developed recent decades. Design and use of these systems were spurred by the progress in instrumentation, communication techniques, and computer technology. In fact, development and use of such systems has become today a standard practice for new engines.

As shown in (Rao, 1996), an advanced monitoring system consists of different components intended to cover all gas turbine subsystems. A diagnostic analysis of registered gas path variables (pressures, temperatures, rotation speeds, fuel consumption, etc.) can be considered as a principal component and integral part of the system. Many different types of gas path performance degradation, such as foreign object damage, fouling, tip rubs, seal wear, and erosion, are known and can be diagnosed. Detailed descriptions of these abrupt faults and gradual deterioration mechanisms can be found, for instance, in (Rao, 1996; Meher-Homji et al., 2001). In addition to the mentioned gas path faults, the analysis of gas path variables (gas path analysis, GPA) also allows detecting sensor malfunctions and wrong operation of a control system (Tsalavoutas et al., 2000). Moreover, this analysis allows estimating main measured engine performances like shaft power, thrust, overall engine efficiency, specific fuel consumption, and compressor surge margin.

The GPA is an area of extensive studies and thousands of published works can be found in this area. Some common observations that follow from the publications and help to explain the structure of the present chapter are given below.

According to known publications, a total diagnostic process usually includes a preliminary stage of feature extraction and three principal stages of monitoring (fault detection), detailed diagnosis (fault localization), and prognosis. Each stage is usually presented by specific algorithms.

The feature extraction means extraction of useful diagnostic information from raw measurement data. This stage includes measurement validation and computing deviations. The deviation of a monitored variable is determined as a discrepancy between a measured value and an engine base-line model. In contrast to the monitored variables themselves that strongly depend on an engine operating mode, the deviations, when properly computed, do not depend on the mode and can be good indicators of an engine health condition. Since the described deviations are input parameters to all diagnostic algorithms, close attention should be paid to the issue of the accuracy of the base-line model and deviations. Some interesting studies, for instance, (Mesbahi et al., 2001; Fast et al., 2009), completely devoted to deviation computation were performed the last decade. One of focuses of the present chapter is on this issue as well.

Among GPA techniques, the fault localization algorithms may be considered as the most important and sophisticated. They involve different mathematical gas turbine models to describe possible faults. In spite of the availability of recorded data, the models are required because real gas turbine faults occur rarely. Recorded data are sufficient to describe only some intensive and practically permanent deterioration mechanisms, such as compressor fouling and erosion. The compressor fouling is the most common cause of the deterioration of stationary gas turbines; its impact on gas turbine performance is well described, see (Meher-Homji et al., 2001). In helicopter engines, compressor airfoils are often affected by erosion because of dust and sand in the sucked air (Meher-Homji et al., 2001).

The models connect faults of different engine components with the corresponding changes of monitored variables, assisting with fault description. Among fault simulation tools, a nonlinear thermodynamic model is of utmost importance and, with this model, many other particular models can be created. Its description includes mass, energy, and momentum conservation laws and requires detailed knowledge of the gas turbine under analysis. The model can be classified as physics-based and presents a complex software package. Such sophisticated models have been used in gas turbine diagnostics since the works of Saravanamuttoo H.I.H. (Saravanamuttoo et al, 1983). The last two decades, the use of these models instead of less exact linear models has become a standard practice.

All fault localization methods can be broken down into two main approaches. The first approach is based on the pattern recognition theory while the second approach applies system identification methods. Reliable fault localization presents a complicated recognition problem because many negative factors impede correct diagnostic decisions. The main factors affecting diagnosis accuracy are a) fault variety and rare occurrence of the same fault, b) inadequacy of the used engine models, c) dependence of fault manifestations on engine operating conditions and engine-to-engine differences, d) sensor noise and malfunctions, and e) control system inaccuracy and possible malfunction. Advances of computer technology have inspired the application of pattern recognition techniques for gas turbine diagnosis. A lot of applications can be found in literature including, but not limited to, Artificial Neural Networks (Roemer et al, 2000; Ogaji et al, 2003; Volponi et al., 2003; Sampath et al., 2006; Butler et al., 2006; Roemer et al, 2000; Greitzer et al, 1999; Romessis et al, 2006; Loboda, Yepifanov et al, 2007; Loboda, Yepifanov et al, 2006), Genetic Algorithms (Sampath et al., 2006), Support Vector Machines (Butler et al., 2006), Correspondence and Discrimination Analysis (Pipe, 1987), and Bayesian Approach (Romessis et al., 2006; Loboda & Yepifanov, 2006). Regardless of applied technique, a fault classification is an integral part of a fault recognition process. Its accuracy is a crucial factor for a success of diagnostic analysis that is why the classification has to be as much close to reality as possible. Since the information about real faults is accumulated along with the time of engine fleet usage, the advantages of the recognition-based approach increase correspondingly.

The diagnostic algorithms based on gas turbine model identification constitute the second approach (Volponi et al, 2003; Tsalavoutas et al, 2000; MacIsaac & Muir, 1991; Benvenuti, 2001; Tsalavoutas et al, 2000; Aretakis et al., 2003; Doel, 2003). The researchers apply different mathematical methods, for instance, Kalman filter (Volponi et al., 2003) and weighted-least-squares (Doel, 2003). Aretakis et al. (Aretakis et al., 2003) use a combinatorial approach in order to get the estimations when input information is limited. When the researchers have in their disposal the data registered through a prolonged period, they calculate successive estimations and analyze them in time (Tsalavoutas et al., 2000). The identification represents an effective technique of model accuracy enhancement. During the identification such fault parameters are determined which minimize the distance between simulated gas path variables and measured ones. Besides the better model accuracy, the simplification of a diagnostic process is provided because the found estimations of the fault parameters contain information of a current technical state of the components. A final diagnostic decision is made without restrictions imposed by a rigid classification. This is a main advantage of the approach.

Although two described approaches are applied first of all for the fault localization, it is easy to show that they can be extended on the other to stages of the diagnostic process. Thus, all GPA methods can be realized through both pattern recognition and system identification techniques. The combination of the approaches is also possible.

Among last trends in the area of the GPA it is also worth to mention the transition from the option of conventional one-point diagnosis (one operating steady state point considered) to the options of multi-point diagnosis (multiple operating points) (Kamboukos & Mathioudakis, 2006) and to the diagnosis under transient operating conditions (Turney & Halasz,1993; Ogaji et al., 2003). To characterize diagnostic efficiency, probabilities of correct and wrong diagnosis united in a so-called confusion matrix (Davison & Bird, 2008; Butler et al., 2006) are widely applied now.

Thus, the gas path analysis can be recognized as a developed area of common gas turbine diagnostics. The GPA embraces different stages, approaches, options, and methods. Total number of algorithms and their variations is very great. Nevertheless, in this great variety of algorithms and publications it is difficult to find clear recommendations on how to design a new monitoring system. This area does not seem to be sufficiently systematized and generalized.

There are advantages from the monitoring system application since the stages of engine testing and production, and it is important that the system be developed as soon as possible. In order to design an effective system in short time, the designer needs clear instructions on how to choose a system structure and how to tailor each system algorithm. That is why the investigations in the area of gas turbine diagnostics should take into consideration real diagnostic conditions as much as possible and should focus on practical recommendations for the designer.

The present chapter focuses on reliability of gas path diagnosis. To enhance overall reliability, every particular problem of a total diagnosis process should be solved as exactly as possible. In the chapter these problems are considered consequently and new solutions are proposed to reduce the gap between simulated diagnostic process and real engine maintenance conditions. The principles are formulated and practical recommendations are given to develop an effective condition monitoring system. The chapter is structured as follows: thermodynamic models, data validation and tracking the deviations, fault classification, fault recognition techniques, multi-point diagnosis, diagnosis under transient conditions, and system identification techniques.

2. Thermodynamic models

A nonlinear thermodynamic model of steady state operation can be characterized as component-based because each gas turbine component is presented in this model by its full manufacture performance map. The model is described by the following structural formula

$$\vec{Y} = F(\vec{U}, \vec{\Theta}).\tag{1}$$

The model takes into account the influence of an operating point (mode) on monitored variables $\vec{Y}$ through the vector $\vec{U}$ of operating conditions (control variables and ambient air parameters). Gas turbine health condition is reflected by means of a vector $\vec{\Theta}$ of fault parameters. As shown in Fig.1, these parameters shift a little the performance maps of engine components (compressors, combustor, turbines, and other devices) and in this way allow simulating different deterioration mechanisms of varying severity.

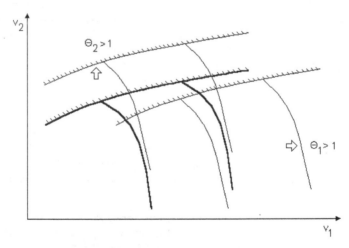

Fig. 1. Component map shifting by the fault parameters (v_1, v_2 - component performances)

The thermodynamic model is a typical physics-based model due to objective physical principles realized. The model has a capacity to simulate baseline engine behaviour. Additionally, since the fault parameters change the component performances involving in the calculations, the model is capable to reflect different types of gas turbine degradation. From the mathematical point of view, equation (1) is a result of the solving a system of nonlinear algebraic equations reflecting mass and energy balance at steady states. All engine components should be thoroughly described to form the equations therefore the model presents complex software including dozens subroutines.

The thermodynamic model can be tailored to real data by means of system identification techniques. As a result, the dependency between variables $\vec{U}$ and $\vec{Y}$ will be exact enough. Nevertheless, it is much more difficult to fit the dependency between Θ and $\vec{Y}$ because empiric information on different faults is not available. The study (Loboda & Yepifanov, 2010) shows that differences between real and simulated faults can be visible.

Software of the nonlinear thermodynamic model allows calculating a matrix of fault influence coefficients H for a linear model

$$\delta\vec{Y} = H\delta\vec{\Theta}.\qquad (2)$$

The linear model computes a vector of relative deviations $\delta\vec{Y}$ induced by small changes of fault parameters $\delta\vec{\Theta}$ at a fixed operating condition. For the changes $\delta\Theta$ typical for real faults, linearization errors are not too great. They seem to be smaller than the mentioned inadequacy of the thermodynamic model in describing the dependency $\vec{Y} = f(\Theta)$. That is why, the linear model can be successfully applied for fault simulation. Additionally, it is useful for analytical analysis of complex diagnostic issues. Consequently, we can state that the model will remain important in gas turbine diagnostics.

If the nonlinear thermodynamic model for steady states is available (static model), a dynamic model can be developed with less efforts. Since transients provide more information than steady states and transient analysis allows continuous diagnosis, the dynamic gas turbine model is in increasing demand. As in the case of the static model, the dynamic model describes how the monitored variables $\vec{Y}$ depend on the quantities $\vec{U}$ and Θ. However, the vector $\vec{U}$ is given as a function of time and a time variable t is also added as an independent argument.

Given the above explanations, the dynamic model is presented by a structural expression

$$\vec{Y} = F(\vec{U}(t), \vec{\Theta}, t).\qquad (3)$$

A separate influence of variable t is explained by inertia of gas turbine rotors, moving gas, and heat interchange processes. Mathematically, the dynamic model is a system of differential equation including time-derivatives and, for each time step, the solution represents a quasi-steady state operating point. Right parts of the differential equations are calculated through the algebraic equation system, which is similar to the system of the static model. That is why, the dynamic model includes the majority of the static model subroutines and these two models tend to form common software package.

Athough the described models are sufficient to simulate a healthy condition and possible faults of gas turbines, the design of monitoring systems cannot be based only on these simulation tools. Models are not always adequate enough and every possibility to make them more accurate comparing with real data should be used. Additionally, diagnostic analysis of recorded data gives new information about possible faults. Although direct tracking time plots of measurements can provide some useful information about gas path degradation and sensor malfunctions, deviations of monitored variables from their baseline values provide more diagnostic information.

3. Data validation and computing the deviations

A systematic change of the deviations is induced by engine degradation, for instance, compressor fouling. Noticeable random errors are also added because of high sensitivity of the deviations to sensor malfunctions and baseline value inaccuracy. As mentioned in section 1, close attention should be paid to deviation accuracy. The ways to make the deviations more accurate are considered below.

3.1 Deviations and sensor malfunctions

Typically, historical engine sensor data previously filtered, averaged, and periodically recorded at steady states are used for gas turbine monitoring and diagnosis. The parameters recorded in the same time include measured values $\vec{U}_m^*$ and Y^* of engine operational conditions and monitored variables correspondingly. For monitored variables $Y_{i,i=1,m}$ the corresponding deviations

$$\delta Y_i^* = \frac{Y_i^* - Y_{0i}(\vec{U}_m^*)}{Y_{0i}(\vec{U}_m^*)} \qquad (4)$$

are computed as relative differences between measured values Y_i^* and engine baseline values $Y_{0i}(\vec{U}_m^*)$. The baseline values are written here as function because a healthy engine operation depends on operating conditions. A totality $\vec{Y}_0(\vec{U}_m)$ of baseline functions is usually called a baseline model.

Figure 2 illustrates the deviations computed for two monitored variables of a gas turbine power plant for natural gas pipelines (Let us call this plant as GT1). It may be concluded about the behavior of the deviations that a compressor washing (t = 7970h) as well as previous and subsequent compressor fouling periods are well-distinguishable. However, the fluctuations are still significant here and capable to mask degradation effects.

As follows from equation (4), the deviation errors can be induced by both malfunctions of sensors and inadequacy of the baseline model. Consequently, it is important to look for and exclude erroneous recorded data as well as to enhance the model. Let us first consider how to detect and identify sensor malfunction cases in recorded data. Various graphical tools help to solve this problem. Deviation plots are very good in malfunction detection but the found deviation anomalies are not always sufficient to determine the anomaly cause: sensor malfunction or model inadequacy. That is why additional plots and theoretical analysis are utilized to identify the anomaly.

Let us handle the problem of sensor malfunctions by the example of EGT measurements. The availability of parallel measurements by a suite of thermocouple probes installed in the same engine station gives us new possibilities of thermocouple malfunction detection by means of deviation analysis. If we choose the same baseline model arguments and the same data to compute model coefficients (these data are called a reference set), the errors related with the model will be approximately equal in deviations of all particular probes. That is why, the differences between deviations of one probe and deviations of the other probes can denote probable errors and faults of this probe. In the synchronous deviation curves constructed versus an engine operation time, such differences will be well visible.

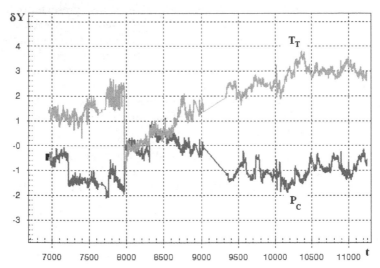

Fig. 2. GT1 deviations δY , % vs. operation time t, hours
(T_T - exhaust gas temperature, P_C - compressor pressure)

Direct analysis of thermocouple probe measurements can be useful as well. To this effect, synchronous plots for all particular probes are constructed vs. the operation time. Engine operating conditions change from one time point to another and this explains common temporal changes of the curves. Anomalies in behaviour of a particular probe can confirm a probe's malfunction. Synchronized perturbations in curves of some probes may be the result of a real temperature profile distortion because of a hot section problem.

A gas turbine driver for an electric generator (let us call it GT2) has been chosen as a test case to analyse possible malfunctions of EGT thermocouple probes (Loboda, Feldshteyn et al, 2009). Figure 3 (a) shows EGT deviations for 5 probes and for the temperature averaged for 11 different probes. It is known that the washings took place at the time points t = 803, 1916, 3098, and 4317. As can be seen, deviation plots reflect in a variable manner the influence of the fouling and washings. The deviation dTt_{med} does it better than deviations of particular probes. Among deviations dTt_i, quantities dTt_5 and dTt_6 , for example, have almost the same diagnostic quality as dTt_{med}, while quantities dTt_1 and dTt_2 are of little quality. Such differences can be partly explained by variations in probe accuracy and reliability. For example, elevated random errors of the deviations dTt_1 and dTt_2 over the whole analyzed period can be induced by greater noise of the first and second EGT probes. The dTt_1 fluctuations in the time interval 1900 - 2600 are probably results of frequent incipient faults of the first probe. However, the shifts of the deviations dTt_1, dTt_2, and dTt_6 around the point t = 3351 present the most interest for the current analysis. The shifts look like a washing result but they have opposite directions.

Parallel plots of probe measurements themselves shown in Fig. 3 (b) confirm the anomalies in recorded data. Small synchronized shifts can be seen here. It is visible in the graph that probes 7 and 8 are synchronously displaced by about 10 degrees during two time intervals t = 962.5-966.5 and t = 971.5-972.5. Additionally, the same measurement increase is observable in the probe 1 curve at time t = 971.5-972.5. In this way, unlike independent outliers found beforehand (Loboda, Feldshteyn et al., 2009), the considered case presents correlated shifts in data of some probes and therefore is more complicated. Two explanations can be proposed for this case. The first of them is related with any common problem of the measurement system that affects some probes and alters their data. So, the outliers can be classified as measurement errors. The second supposes that the measurements are correct but a real EGT profile has been changed in the noted time points. It can be possible because there is no information that EGT and PTT probe profiles should be absolutely stable during engine operation. The available data are not sufficient to give a unique explanation; more recorded data should be attracted.

In addition to the case of thermocouples considered before, many other cases of abnormal sensor data behaviour have been detected and interpreted. The reader can find them in (Loboda, Feldshteyn et al., 2009; Loboda et al., 2004; Loboda, Yepifanov et al., 2009).

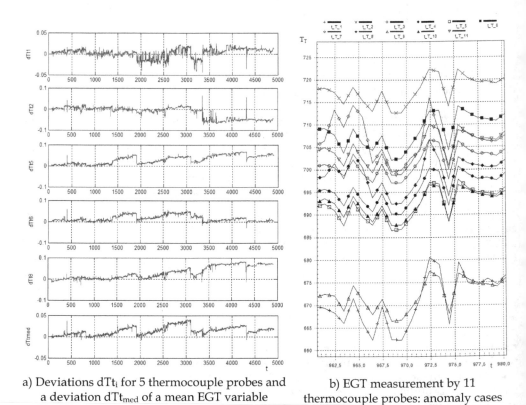

a) Deviations dTt$_i$ for 5 thermocouple probes and a deviation dTt$_{med}$ of a mean EGT variable

b) EGT measurement by 11 thermocouple probes: anomaly cases

Fig. 3. EGT deviations and temperatures themselves (GT2)

3.2 Deviations and baseline model adequacy

In addition to rectification of recorded data, baseline model improvement provides further enhancement of deviation accuracy. Adequacy of this model depends on a correct choice of three elements: model arguments (elements of the vector $\vec{U}$), type of an approximating function, and reference set. Many variations of these factors have been proposed, realized and compared (Loboda et al., 2004; Loboda, Yepifanov et al., 2009) in order to choose the best choice of each factor. The criterion to compare the choices was an integral quality of the corresponding deviations. Table 1 reflects the considered cases.

Summarizing the results of all proposals to make the baseline model more adequate, we draw two general conclusions. First, the majority of the proposed choices improved the deviations; however, the improvement was always very limited. In other words, many choices yielded practically the same equal deviation accuracy about 0.5±%. This means that principally new choices should be proposed and considered to significantly enhance the deviations. Second, not all real operating conditions were included as arguments of the baseline model. Some variables of real operating conditions are not always measured or recorded, for example, inlet air humidity, air bleeding and bypass valves' positions, and engine box temperature. Since such variables exert influence upon a real engine and its monitored variables but are not taken into consideration in the baseline model, the corresponding deviation errors take place. A similar negative effect can occur if sensor systematic error changes along with operation time.

Model arguments	Approximating function	Reference set
• Selection of the best power set parameter • Selection of the best ambient air parameters • Including additional arguments (e.g. humidity parameter)	• Second order polynomials determined by the singular value decomposition (SVD) method • Full second order polynomials determined by the least square method (LSM) • Full third order polynomials determined by the least square method (LSM) • Neural networks, in particular, multilayer perceptron	• Reference set of different volume • Reference set computed by the thermodynamic model • Reference set of great volume for a model of a degraded engine. This model is simply converted into the necessary baseline model

Table 1. Choices of the baseline model elements

Thus, practically all ways to enhance the baseline model with available real data and thermodynamic model have been examined. The achieved deviation accuracy is not too low.

That is why, new considerable accuracy enhancement will not come easily. It is necessary to find new operating condition variables, to determine their influence on monitored variables, and to make the baseline model more accurate. It seems to us that this will require a hard combined work of engine and sensor designers, maintenance personnel, and diagnosticians.

As a small part of this work, the next subsection analyzes possible sources of deviation errors. This analysis firstly performed in (Loboda, 2011).

3.3 Theoretical analysis of possible errors in real deviations

The analysis is based on expression (4). For a monitored variable Y, this expression can be rewritten as

$$\delta Y^* = \frac{Y^*}{\hat{Y}_0(\vec{U}_m^*)} - 1 . \tag{5}$$

This equation shows that inaccuracy of the deviation is completely determined by errors in a term $Y^*/\hat{Y}_0(\vec{U}_m^*)$, in which $\hat{Y}_0$ denotes an estimation of the baseline function for the variable Y. It is shown below that these errors can be divided into four types.

The measurement Y^* differs from a true value Y by an error E_Y called in this paper as a Type I error. In its turn, the true value depends on a vector $\vec{U}$ of real operating conditions and on engine health conditions given by the vector $\Delta\vec{\Theta}$. As a consequence, the value Y^* can be determined as

$$Y^* = Y(\vec{U},\Delta\vec{\Theta}) + E_Y(\vec{U},\Delta\vec{\Theta}) . \tag{6}$$

The error E_Y is defined here as a function because, in general, measurement errors may depend on the value Y and, consequently, on the variables $\vec{U}$ and $\Delta\vec{\Theta}$.

One more obvious cause of the deviation inaccuracy is related with measurement errors in operating conditions presented in equation (5) by the vector $\vec{U}_m^*$. Given a vector of measurement errors $\vec{E}_{Um}$, which presents Type II errors, the measured operating conditions are written as

$$\vec{U}_m^* = \vec{U}_m + \vec{E}_{Um} . \tag{7}$$

The next error type (Type III) is also related to engine operating conditions however it is not so evident. The point is that not all real operating conditions denominated in the present paper by a $[(n+k)\times1]$-vector $\vec{U}$ can be included as arguments of the baseline function. Some variables of real operating conditions are not always measured or recorded, for example, inlet air humidity, air bleeding and bypass valves' positions, and engine box temperature. Let us unite all these additional variables in a $(k\times1)$-vector $\vec{E}_U$. Since such variables exert influence upon a real engine and its measured variable Y^* but are not taken into consideration in the baseline function $\hat{Y}_0(\vec{U}_m^*)$, the corresponding deviation errors take place. A similar negative effect can occur if sensor systematic error changes in time. Given that $\vec{U} = \vec{U}_m \cup \vec{E}_U$, the vector $\vec{U}_m$ can be given by $\vec{U}_m = \vec{U} \setminus \vec{E}_U$ and the equation (9) is converted to a form

$$\vec{U}_m^* = \vec{U} \setminus \vec{E}_U + \vec{E}_{Um} . \tag{8}$$

Apart from the described errors related to the arguments of the function $\hat{Y}_0(\vec{U}_m^*)$, the function has a proper error E_{Y_0} (Type IV error). It can result from such factors as a systematic error in measurements of the variable Y, inadequate function type, improper algorithm for estimating function's coefficient, errors in the reference set, limited volume of the set data, and influence of engine deterioration on these data. Given E_{Y_0} and a true function Y_0, the function estimation $\hat{Y}_0$ can be written as

$$\hat{Y}_0(\vec{U}_m^*) = Y_0(\vec{U}_m^*) + E_{Y_0}(\vec{U}_m^*) . \tag{9}$$

Let us now substitute equations (6), (8), and (9) into expression (5). As a result, the deviation δY^* is written as

$$\delta Y^* = \frac{Y(\vec{U}, \Delta\vec{\Theta}) + E_Y(\vec{U}_m^* - \vec{E}_{Um} + \vec{E}_U, \Delta\vec{\Theta})}{Y_0(\vec{U} \setminus \vec{E}_U + \vec{E}_{Um}) + E_{Y_0}(\vec{U}_m^*)} - 1 . \tag{10}$$

A dependency $E_Y(\vec{U}_m^* - \vec{E}_{Um} + \vec{E}_U, \Delta\vec{\Theta})$ in this expression can be simplified because of the following reasons: a) $E_Y \ll Y$, b) $\left\|\vec{E}_{Um}\right\| \ll \left\|\vec{U}_m^*\right\|$, c) The influence of $\vec{E}_U$ and $\Delta\vec{\Theta}$ on Y and, consequently, on E_Y is significantly smaller then the influence of $\vec{U}_m^*$. Taking into account the considerations made, we arrive to a final expression for the deviation

$$\delta Y^* = \frac{Y(\vec{U}, \Delta\vec{\Theta}) + E_Y(\vec{U}_m^*)}{Y_0(\vec{U} \setminus \vec{E}_U + \vec{E}_{Um}) + E_{Y_0}(\vec{U}_m^*)} - 1 . \tag{11}$$

This expression includes four error types introduced above, namely $E_Y, \vec{E}_{Um}, \vec{E}_U$, and E_{Y_0} . Let us now analyze how each error can influence on inaccuracy of the deviation δY^* . This analysis is performed under the following assumptions. First, the same sensors were employed to measure currently analyzed values Y^* and U_m^* as well as the reference set data. Second, gross errors have been filtered out. Third, a systematic error and distribution of random errors in Y^* and $\vec{U}_m^*$ do not depend on engine operating time.

Type I error. Since the sensor performance is invariable, every systematic change of the error E_Y will be accompanied by the same change in E_{Y_0} . As a consequence, accuracy of the deviation δY^* will not be affected by the systematic component of E_Y . As to the random component, it is usually given by the multidimensional Gaussian distribution. That is why, the corresponding error component in the deviations δY^* can also be described by this distribution.

Type II errors. It is easy to show that the system component of the errors $\vec{E}_{Um}$ cannot influence a lot the deviation δY^*. As to the random component, it can be described by the multidimensional Gaussian distribution, as in the case of the monitored variables Y. Because every change of the arguments $\vec{U}_m^*$ has an influence on baseline values of all monitored variables, their baseline values $\hat{Y}_0$ and, consequently, deviations δY^* have correlation. Thus, random errors of operating conditions induce correlated deviation errors that cannot be described by the multidimensional Gaussian distribution. It is very likely that permanent noise with a small scatter observed in Fig.2 results from the errors of Type I and Type II.

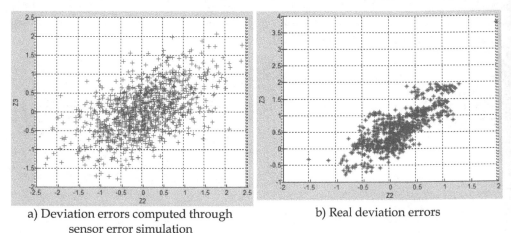

a) Deviation errors computed through   b) Real deviation errors
sensor error simulation

Fig. 4. Simulated and real deviation errors of the GT1 (Z2 – normalized deviation of the EGT; Z3 - normalized deviation of the power turbine temperature)

Type III error. Presence of such an error has been confirmed after analyzing all other error types. This error occurs because the additional operating conditions $\vec{E}_U$ do not change baseline function but exert influence on a real engine and, accordingly, on all variables Y. For this reason, any change of $\vec{E}_U$ can induce synchronous errors of the deviations δY^* of all monitored variables. It is very likely that most deviation fluctuations in Fig.2 origin from the Type III errors.

Type IV error. This error varies in time along with changes in the operating conditions $\vec{U}_m^*$ producing perturbations in the deviation variable δY^*. The perturbations can be both independent and correlated depending on particular causes of the error E_{Y_0}. Although the baseline function adequacy is a challenge, the error can be reduced to an acceptable level by applying a proper function type and using a representative reference set.

Distributions of simulated and real deviation errors are shown in Fig.4. Normalized deviations

$$Z^* = \delta Y^* / a_Y \qquad (12)$$

are presented here. A parameter a_Y denotes an amplitude of random errors in the deviation variable δY^*. Such normalization simplifies fault class description and enhances diagnosis reliability. Diagram (a) illustrates the deviation errors simulated through the multidimensional Gaussian distribution of sensor errors (Type I and Type II errors). Such simulation is traditionally applied in gas turbine fault recognition algorithms. Diagram (b) shows the errors extracted from real data-based deviations. Both diagrams show visible error correlation between the presented deviations. But there are visible differences as well: the distribution of real deviation errors is less regular. Not taken into consideration in fault recognition algorithms, these differences can affect the reliability of gas turbine diagnosis. Therefore, to make the diagnosis more reliable, simulated noise should be as close as possible to real errors.

4. Classification for fault recognition algorithms

As mentioned in the introduction, mathematical models are widely used in gas turbine diagnostics to describe engine performance degradation and faults and the deviations are employed to reveal the degradation influence. For this reason, a classification for fault recognition algorithms is usually formed in the space of the deviations with the use of nonlinear or linear gas turbine models.

4.1 Fault classification in the deviation space

For the purposes of diagnosis, existing variety of engine faults should be broken down into a limited number of classes. The hypothesis commonly used in the pattern recognition theory states that a recognized object can belong only to one of q classes

$$D_1, D_2, ..., D_q \tag{13}$$

that are set before recognition itself. This assumption is also accepted in gas turbine diagnostics. As a rule, each fault class corresponds to one engine component and is described by its fault parameters. If we change only one fault parameter, a single fault class is formed whereas a multiple fault class is created by an independent variation of two and more fault parameters.

As mentioned in the introduction, the deviations are good indicators of engine faults. That is why the normalized deviations Z computed for m available monitored variables form an appropriate space to recognize the faults. On the basis of the above considerations, a recognition space (diagnostic space) is formed on the basis of the vector $\vec{Z}$ ($m \times 1$) that unites elemental normalized deviations. One value of the vector $\vec{Z}^*$ computed with the measurements $\vec{Y}^*$ presents a pattern to be recognized.

Some recognition techniques, for example, the Bayesian approach, need a probabilistic description of the used classification. In this case each fault class D_j should be described by its probability density function $f(\vec{Z}^* / D_j)$. The difficulty of this approach is related with density functions themselves because it is a principal problem of mathematical statistics to assess them. That is why the probabilistic description can be realized only for a simplified fault classes.

More recognition techniques use a statistical classification description. In this case the fault classes are given by samples of patterns, namely vectors Z^*. In this way, a whole fault classification is a union of pattern samples of all classes. Apart from the simplification of a class formation process, the replacement of the density functions by pattern samples allows creating more complex fault classes only on the basis of real data. However, gas turbine faults are still often simulated mathematically because of rare appearance of real faults and high costs of physical fault simulation.

The deviations, induced by the faults that are embedded into the thermodynamic model via a change $\Delta\vec{\Theta}$, can be computed according to a formula

$$Z_i = \frac{Y_i(\vec{U},\vec{\Theta}_0 + \Delta\vec{\Theta}) - Y_i(\vec{U},\vec{\Theta}_0)}{Y_i(\vec{U},\vec{\Theta}_0)a_{Yi}} + \varepsilon_i, i = 1, m .$$ (14)

The vector $\vec{\Theta}_0$ corresponds here to a healthy engine. Random errors ε_i make deviations more realistic. They can be added directly to systematic parts of the deviations or can be introduced through the simulation of random measurement errors in Y_i and $\vec{U}$. The deviation vectors Z^* (patterns) for faults of different type and severity are generated by the model through changing a structure and length of the vector $\Delta\vec{\Theta}$. The resulting totality **Zl*** of all classification's patterns is typically called a learning set because it is applied to train the used recognition technique, for example, a neural network.

There is a common statistical rule that a function determined on one portion of the random data should be tested on another. Consequently, to verify the technique trained on the learning set, we need one more set. The necessary set **Zv***, called a validation set, is created in the same way as the set **Zl***. The only exception is that different series of the random numbers are involved in the calculations of the fault severity and errors in the deviations. A class of every pattern of the validation set is beforehand known. Therefore, applying the trained technique to this set, we can compare the diagnosis with a known class and compute a vector $\vec{P}$ of true diagnosis probabilities and an averaged probability $\overline{P}$. These probabilities quantify class distinquishability and engine diagnosability and are good criteria to tailor and compare recognition techniques.

The patterns of the learning and validation sets described above are generated at a fixed operating mode given by a constant vector $\vec{U}$. Such a classification is intended for diagnosis at the same mode. The principal to make the classification and fault recognition more universal is described below.

4.2 Universal fault classification

The principle of a universal fault classification had been proposed and investigated in [Loboda & Feldshteyn, 2007; Loboda & Yepifanov, 2010). During the diagnosis at different operating modes, it has been found that class presentation in the diagnostic space Z is not strongly dependent on a mode change. Therefore we intended to draw up the classification that would be independent from operational conditions. The classification has been created by incorporating patterns from different steady states into every class. In this case, a region occupied by a class is more diffused inducing greater class intersection. This objectively leads to additional losses of the diagnosis reliability but the investigations have shown that these losses are insignificant. Such new classification was compared with a conventional classification for one operating mode. The comparison was made under different diagnostic conditions (different engines, steady state operating conditions and fault class types). Additionally, the comparison was performed under transient operating conditions. The resulting losses did not exceed 2%. Thus, the universal classification does not significantly reduce the diagnosis reliability level. On the other hand, the suggested classification drastically simplifies the gas turbine diagnosis because it is formed once and used later without changes to diagnose an engine at any operating mode. Therefore, the diagnostic algorithms based on the universal fault classification can be successfully implemented in real condition monitoring systems.

Another way to enhance a convenient classification is related with the idea to embed a real fault class into model based classification. The idea of such a mixed classification has been proposed and validated in paper (Loboda & Yepifanov, 2010).

4.3 Mixed fault classification

Since model errors, which can be significant, are transmitted to the model-based classification, the idea appears to make the description of some classes more accurate using real data. Such a mixed fault classification will incorporate both model-based and data-driven fault classes. The classification will combine a profound common diagnosis with a higher diagnostic accuracy for the data-driven classes. To support the idea, a data-driven class of the fouling based on real fouling data has been created and incorporated into the model-based classification. The resulting mixed classification and a convinient model-based classification were embedded into a diagnostic algorithm. It was found that the application of a model based classification to real data influenced by compressor fouling causes severe diagnosis errors of over 30 per cent. However, the switch to the mixed classification results in a decrease of error of up to 3 percent only.

The next way to make the classification more realistic consists in the insertion of real deviation errors into the description of model based classes proposed in (Loboda, 2011).

4.4 Classification with more realistic deviation errors

The most of researchers also take into account random errors in the monitored variables and operating conditions applying the Gaussian distribution to that end. However, as shown in section 3, the difference between such traditionally simulated errors and real deviation errors can be significant. That is why it is proposed to draw a noise part from the deviations and integrate it into the description of simulated fault classes. Three alternative schemes, two existing and one new, of deviation error representation in diagnostic algorithms have been realized. They were compared with the use of probability of correct diagnosis $\bar{P}$. Figure 5 illustrates the fault classification with simulated measurement errors like in Fig.4 (a).

Preliminary calculations have shown that the distinguishability of fault classes can change by up to 6% when real errors are replaced by simulated errors. Thus, the diagnostic performance estimated with simulated noise can be inaccurate. The case was also investigated when the errors for the learning and validation set were extracted from different time portions of real data. The loss of diagnosability for this case was found drastic: from $\bar{P}$ = 90% - 94% in the previous cases to $\bar{P}$ = 59%. It has happened because real deviation errors included into the validation set increased a lot in comparison with the learning set errors. The increase of the errors occurred because the baseline model is adequate on the reference set data but loses its accuracy on the subsequently recorded data. Such a problem seems to be very probable in real diagnosis and we should be careful to avoid or mitigate it.

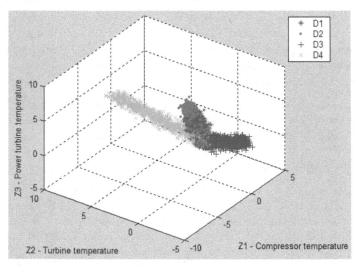

Fig. 5. 3D plot of four fault classes with simulated sensor errors

Although the proposed scheme is more realistic, it cannot automatically replace existing noise simulation modes. This new scheme is more complex for realization. Additionally, it needs both the thermodynamic model and extensive real data, two things rarely available together. In this way, the proposed scheme of deviation error representation can rather be recommended for a final precise estimation of gas turbine diagnosability.

Thus, we completed the analysis of different improvements of a convenient fault classification. Let us now consider the problem of choosing a recognition technique.

5. Recognition technique selection

On basis of the probabilistic indices $\vec{P}$ and $\bar{P}$, three recognition techniques are optimized and compared in paper (Loboda & Yepifanov, 2006) in order to choose the best one and give recommendations on its practical use. Two techniques, the Bayesian approach and multilayer perceptron (type of neural networks), have shown similarly good results. The difference of the probabilities $\bar{P}$ estimated for each technique was only about 0.6% in different conditions of technique application.

To continue the comparison of recognition techniques, paper (Loboda, Feldshteyn et al., 2011) compares two network types: a multilayer perceptron (MLP) and a radial basis network (RBN). To draw firm conclusions on the networks' applicability, comparative calculations were repeated for different variations of diagnostic conditions. In particular, two different engines were chosen as test cases. The comparison results are shown in Table 2. It can be seen that the differences between the techniques are very smal. On average for all cases presented in the table for the GT1, the RBN gains 0.0009 (0.09%) only.

Class type	Network type	Mode 1		Mode 2	
		Basic node numbers	Enlarged node numbers	Basic node numbers	Enlarged node numbers
Singular		Case 1	Case 2	Case 3	Case 4
	MLP	0.8157	0.8184	0.8031	0.8059
	RBN	0.8115	0.8186	0.8009	0.8058
Multiple		Case 5	Case 6	Case 7	Case 8
	MLP	0.8738	0.8765	0.8660	0.8686
	RBN	0.8745	0.8783	0.8663	0.8701

Table 2. Probabilities $\bar{P}$ for the networks compared on GT1 data

The comparison was repeated for an aircraft turbofan engine denoted as GT3. The corresponding average probability increment was found to be 0.0028 (0.28%). In this way, an advantage of the radial basis network in the application to the analyzed turbofan seems to be a little more notable than in the case of the industrial gas turbine. By way of summing up the comparison results, the conclusion is that the radial basis network is a little more accurate than the perceptron, however the difference can be considered as insignificant.

The comparison of recognition techniques has been completed in paper (Estrada Moreno & Loboda, 2011), in which the MLP and probabilistic neural network (PNN) are compared. The comparison under different diagnostic conditions has revealed that the diagnosis by the PNN is less reliable. However, the averaged difference of the probability is not greater than 0.5%. Once more we can state that the compared techniques are practically equal in accuracy.

The fact that the fault recognition performances of four different recognition techniques, namely, Bayessian approach, MLP, RBN, and PNN, are very close is worthy of some discussion. What explanation can be provided for this? We believe that all four techniques are sophisticated enough and are well suited for solving this specific problem – gas turbine fault recognition. No one of these techniques can further enhance diagnostic accuracy because the accuracy achieved is near the theoretical accuracy level that is inherent to the solving problem: gas turbine fault recognition with the given classification. Following this idea, we suppose that within the approach used and the classification accepted no other recognition technique will be capable to considerably enhance diagnostic accuracy. Instead, all efforts should be made to reduce fault class intersections, for example, by reducing measurement inaccuracy, installing more sensors in the gas path, and decreasing deviation errors. The options of multipoint diagnosis and diagnosis during transient operation will also result in a higher diagnostic accuracy.

6. Multipoint diagnosis and diagnosis at transients

The options of one-point diagnosis, multipoint diagnosis, and diagnosis at transients have been thoroughly studied and compared in (Loboda & Feldshteyn, 2007). The same probabilistic criteria $\vec{P}$ and $\bar{P}$ were used to compare the recognition techniques. The multipoint diagnosis means that measurements from different operating points (modes) are united to make a single diagnosis. All recognition techniques previously described in the application within the one-point option can be applied without principal changes. The dimension of patterns and diagnostic space is only increased because measurements at every operating mode can be considered as new gas turbine measured variables. In this way, a generalized deviation vector $\vec{W}^*$, which unites deviations computed at all considered modes, is now a pattern to be recognised. For the investigated case of the GT3 when the engine has 5 five monitored variables and is diagnosed at 14 modes, the generalized vector $\vec{W}^*$ embraces 5×14=70 elemental deviations. To illustrate the effect of the multipoint option, Table 3 presents the results of the comparison of the one-point and multipoint options. The probability $\bar{P}$ increments contained in the line "Difference" allow to state that a positive effect from using the multipoint option is very significant. A principal part of these increments is explained by so-called averaging effect (Loboda & Feldshteyn, 2007) of the multipoint diagnosis.

From the mathematical point of view, the diagnosis under transient condition is similar to the multipoint diagnosis: every measurement section of a total transient process is considered as a new operating point and the same generalized vector $\vec{W}^*$ is formed. This allows comparing these two options. With the comparison results given in Table 4, one can state that the diagnosis at transients has a stable, although not very high, growth of accuracy relative to the multipoint diagnosis. This growth is probably related to the greater fault influence under dynamic conditions. As the growth is not considerable, the actions of diagnosis at transients and multipoint diagnosis are close. Consequently, the most part of the total accuracy growth at transients relative to the one-point option is produced by the averaging effect mentioned above.

Option	Single fault classification	Multiple fault classification
One-point	0.7316	0.7351
Multipoint	0.8915	0.9444
Difference	0.1599	0.2093

Table 3. Probabilities $\bar{P}$ for the one-point diagnosis and multipoint diagnosis (GT3)

Option	Single fault classification	Multiple fault classification
Multipoint	0.8915	0.9444
Transient	0.9032	0.9561
Difference	0.0117	0.0117

Table 4. Probabilities $\bar{P}$ for the multipoint diagnosis and diagnosis at transients (GT3)

When simulating the diagnostic processes at transients, we were unable to take into account some peculiarities that complicate the real diagnosis at transients, such as a turbine temperature sensor's dynamic error and an unequal dynamic warm-up of rotor and stator parts. That is why our conclusions regarding the diagnosis at transients cannot be considered as the sole argument for choosing a proper diagnostic option.

Thus, the modes and options to enhance the diagnosis methods based on the pattern recognition theory have been analysed in details in sections 4-6. In contrast, the next section gives a general description of the other main diagnostic approach using system identification techniques.

7. Diagnosis with system identification techniques

The gas turbine diagnosis based on system identification means the identification of the thermodynamic models, namely nonlinear static model, linear model, and dynamic model, given by equations (1) - (3). The techniques to identify the nonlinear static model are most widely used nowadays in the GPA. The techniques compute estimates $\hat{\vec{\Theta}}$ as a result of distance minimization between simulated and measured values of the monitored variables. This minimization problem can be written as

$$\hat{\vec{\Theta}} = \arg\min \left\| \vec{Y}^* - \vec{Y}(\vec{U}, \vec{\Theta}) \right\|. \tag{15}$$

The estimates contain information on a current technical state of each engine component. This drastically simplifies a subsequent diagnostic decision. Furthermore, the diagnosis is not limited by a rigid classification as in the case of the pattern recognition-based approach.

Among the system identification techniques applied to diagnose gas turbines, the Kalman filter is by far widely used. The details its application can be found, for example, in (Volponi et al., 2003). The alignment of the estimates is provided by the Kalman filter because its current estimate depends on previous ones.

Other computational scheme is maintained in (Loboda, 2007). Independent estimations are obtained by a special inverse procedure within the multipoint option. With data registered through a prolonged period, successive estimates are computed and analyzed in time to get more reliable results. Within this scheme, a regularizing identification procedure is proposed and verified on simulated and real data in (Loboda et al., 2005). The verification has shown that the regularization of the estimated state parameters makes the identification procedure more stable and reduces an estimation scatter. On the other hand, the regularization shifts mean values of the estimates and should be applied carefully. The values 0.02-0.03 of the regularization parameter were recommended.

Next diagnostic development of the gas turbine identification is presented in (Loboda, 2007). The idea is proposed to develop in the basis of the thermodynamic model a new model that takes gradual engine performance degradation in consideration. Two purposes are achieved identifying such a model. The first purpose consists in creating the model of a gradually degraded engine while the second is to have a baseline function of high accuracy. The idea is verified on maintenance data of the GT1. The comparison of the modified and original identification procedures has shown that the proposed procedure has better properties. Such a model can be widely used in monitoring systems as well.

Another way to get more exact estimates is to use the option of the diagnosis at transients and to identify the dynamic model as shown in (Loboda & Hernandez Gonzalez, 2002).

As shown above, the considered system identification-based approach can be realized for the same options as the pattern recognition-based approach. Additionally, the principal problems of the latter (such as inadequacy of the baseline model and inaccurate fault simulation) are also typical for the system identification-based approach. Thus, they are alternative approaches for all steps of a total fault localization process. Furthermore, it can also be demonstrated that with this approaches the stages of fault detection and prognostics can be realized. In this way, we can consider these two approaches as applicable for all gas path analysis methods.

8. Conclusion

The present chapter is devoted to the enhancement of gas path diagnosis reliability. Different approaches are considered and main trends in gas turbine diagnostics are analyzed by the reviewing multiple literature sources.

It was shown that in many cases such convenient ways to enhance the reliability, as choosing the best approximation function and recognition technique as well as tailoring the function and technique, do not yield significant results nowadays. This happens because of many investigations already conducted in this area.

Some new solutions are proposed in the chapter to reduce the gap between simulated diagnostic process and real engine maintenance conditions. Possible error sources are examined in the chapter and some methods are proposed to enhance the deviation accuracy. In addition, new principles are considered to create a more realistic fault classification, for example, by generating real error distribution.

Among the principal problems to solve in future, insufficient adequacy of the baseline model was detected. The other challenging problem consists in fault simulation inaccuracy.

We hope that the observations made in this chapter and the recommendations drawn will help to design and rapidly tailor new gas turbine health monitoring systems. A long list of references can also be useful for the reader.

9. Acknowledgment

The work has been carried out with the support of the National Polytechnic Institute of Mexico (research project 20113092).

10. References

Aretakis, N.; Mathioudakis, K. & Stamatis, A. (2003). Nonlinear engine component fault diagnosis from a limited number of measurements using a combinatorial approach. *Journal of Engineering for Gas Turbines and Power*, Vol.125, Issue 3, (July 2003), pp. 642-650

Benvenuti, E. (2001). Innovative gas turbine performance diagnostics and hot parts life assessment techniques, *Proceedings of the Thirtieth Turbomachinery Symposium*, pp.23-31, Texas, USA, September 17-20, 2001,Texas A&M University, Houston

Butler, S.W; Pattipati, K.R.; Volponi, A. et al. (2006). An assessment methodology for data-driven and model based techniques for engine health monitoring, *Proceedings of IGTI/ASME Turbo Expo 2006*, 9p., Barcelona, Spain

Davison, C.R. & Bird, J.W. (2008). Review of metrics and assignment if confidence intervals for health management of gas turbine engines, *Proceedings of IGTI/ASME Turbo Expo 2008*, 11p., Berlin, Germany, June 9-13, 2008, ASME Paper GT2008-50849

Doel, D.L. (2003). Interpretation of weighted-least-squares gas path analysis results. *Journal of Engineering for Gas Turbines and Power*, Vol. 125, Issue 3. (July 2003),- pp. 624-633

Fast, M.; Assadi, M.; Pike, A. & Breuhaus, P. (2009). Different condition monitoring models for gas turbines by means of artificial neural networks, *Proceedings of IGTI/ASME Turbo Expo 2009*, 11p., Florida, USA, June 8-12, Orlando, ASME Paper GT2009-59364

Greitzer, F. L.; Kangas, L. J.; Terrones, K. M. et al. (1999) Gas Turbine Engine Health Monitoring and Prognostics, *International Society of Logistics (SOLE) Symposium*, 7p., Las Vegas, Nevada

I. Loboda. (2011). A more realistic presentation of measurement deviation errors in gas turbine diagnostic algorithms. *Aerospace Technics and Technology. Journal: National Aerospace University, Kharkov, Ukraine*, Issue 5, 10p.

Kamboukos, Ph. & Mathioudakis, K. (2006). Multipoint non-linear method for enhanced component and sensor malfunction diagnosis, *Proceedings of IGTI/ASME Turbo Expo 2006*, 9p, Barcelona, Spain, May 8-11

Loboda, I. & Feldshteyn, Ya. (2007). A universal fault classification for gas turbine diagnosis under variable operating conditions, *International Journal of Turbo & Jet Engines*, Vol. 24, No. 1, pp. 11-27, ISSN 0334-0082

Loboda, I. & Hernandez Gonzalez, J. C. (2002). Nonlinear Dynamic Model Identification of Gas Turbine Engine, *Aerospace Technics and Technology. Journal: National Aerospace University, Kharkov, Ukraine*, No. 31, pp. 209 - 211, ISBN 966-7427-08-0, 966-7458-58-X

Loboda, I. & Yepifanov, S. (2006). Gas Turbine Fault Recognition Trustworthiness, *Cientifica*, ESIME-IPN, Mexico, Vol. 10, No. 2, pp. 65-74, ISSN 1665-0654

Loboda, I. & Yepifanov, S. (2010). A Mixed Data-Driven and Model Based Fault Classification for Gas Turbine Diagnosis, *Proceedings of ASME Turbo Expo 2010: International Technical Congress*, 8p., Scotland, UK, June 14-18, Glasgow, ASME Paper No. GT2010-23075

Loboda, I. (2007). Gas turbine diagnostic model identification on maintenance data of great volume, *Aerospace Technics and Technology. Journal: National Aerospace University*, Kharkov, Ukraine, No. 10(46), pp. 198 – 204, ISSN 1727-7337

Loboda, I.; Feldshteyn, Y. & Villarreal González, C.F. (2009). Diagnostic analysis of gas turbine hot section temperature measurements. *Aerospace Technics and Technology. Journal: National Aerospace University, Ukraine*, Issue 6(63), pp.66-79

Loboda, I.; Feldshteyn, Ya. & Ponomaryov, V. (2011). Neural networks for gas turbine fault identification: multilayer perceptron or radial basis network?, *Proceedings of ASME Turbo Expo 2011: International Technical Congress*, 11p., Vancouver, Canada, June 6-10, 2011, ASME Paper No. GT2011-46752

Loboda, I.; Yepifanov, S. & Feldshteyn, Ya. (2004). Deviation problem in gas turbine health monitoring, *Proceedings of IASTED International Conference on Power and Energy Systems*, 6p., Clearwater Beach, Florida, USA

Loboda, I.; Yepifanov, S. & Feldshteyn, Ya. (2007). A generalized fault classification for gas turbine diagnostics on steady states and transients, *Journal of Engineering for Gas Turbines and Power*, Vol. 129, No. 4, pp. 977-985

Loboda, I.; Yepifanov, S. & Feldshteyn, Ya. (2009). Diagnostic analysis of maintenance data of a gas turbine for driving an electric generator, *Proceedings of IGTI/ASME Turbo Expo 2009*, 12p., Florida, USA, June 8-12, Orlando, ASME Paper No. GT2009-60176

Loboda, I.; Zelenskiy, R.; Nerubasskiy, V. & Lopez y Rodriguez, A.R. (2005). Verification of a gas turbine model regularizing identification procedure on simulated and real data, *Memorias del 4to Congreso Internacional de Ingenieria Electromecanica y de Sistemas*, *ESIME, IPN*, 6p., Mexico, November 14-18, ISBN: 970-36-0292-4

MacIsaac, B. D. & Muir, D. F. (1991). Lessons learned in gas turbine performance analysis, *Canadian Gas Association Symposium on Industrial Application of Gas Turbines*, 28p., Canada

Meher-Homji, C.B.; Chaker, M.A. & Motivwala, H.M. (2001). Gas turbine performance deterioration, *Proceedings of Thirtieth Turbomachinery Symposium*, pp.139-175, Texas, USA, September 17-20, 2001,Texas A&M University, Houston

Mesbahi, E.; Assadi, M. et al. (2001). A unique correction tchnique for vaporative gas turbine (EvGT) parameters, *Proceedings of IGTI/ASME Turbo Expo 2001*, 7p., New Orleans, USA, June 4-7, 2001, ASME Paper 2001-GT-0008

Ogaji, S.O.T.; Li, Y. G.; Sampath, S. et al. (2003). Gas path fault diagnosis of a turbofan engine from transient data using artificial neural networks, *Proceedings of IGTI/ASME Turbo Expo 2003*, 10p., Atlanta, Georgia, USA

Pipe K. (1987). Application of Advanced Pattern Recognition Techniques in Machinery Failure Prognosis for Turbomachinery, *Proc. Condition Monitoring 1987 International Conference*, pp. 73-89, British Hydraulic Research Association, UK

Rao, B.K.N. (1996). *Handbook of Condition Monitoring*, Elsevier Advanced Technology, Oxford

Roemer, M.J. & Kacprzynski, G.J. (2000) Advanced diagnostics and prognostics for gas turbine engine risk assessment, *Proceedings of ASME Turbo Expo 2000*, – 10 p., Munich, Germany

Romessis, C. & Mathioudakis, K. (2006). Bayesian network approach for gas path fault diagnosis, *Journal of Engineering for Gas Turbines and Power*, Vol. 128, No. 1, pp. 64-72.

Sampath, S. & Singh, R. (2006). An integrated fault diagnostics model using genetic algorithm and neural networks, *Journal of Engineering for Gas Turbines and Power*, Vol. 128, No. 1, pp. 49-56

Saravanamuttoo, H. I. H. & MacIsaac, B. D. (1983). Thermodynamic models for pipeline gas turbine diagnostics, *ASME Journal of Engineering for Power*, Vol.105, No. 10, pp. 875-884

Strada Moreno R. & Loboda I. (2011). Redes Probabilísticas Enfocadas al Diagnóstico de Turbinas de Gas, *Memorias del 6to Congreso Internacional de Ingeniería Electromecánica y de Sistemas*, 6p., ESIME, IPN, México, D.F., 7-11 noviembre de 2011

Tsalavoutas A., Mathioudakis K., Aretakis N et al. (2000). Combined advanced data analysis method for the constitution of an integrated gas turbine condition monitoring and diagnostic system, *Proceedings of IGTI/ASME Turbo Expo*, 8p., Munich, Germany

Tsalavoutas, A.; Stamatis, A.; Mathioudakis, K. et al. (2000). Identifying Faults in the Variable Geometry System of a Gas Turbine Compressor, *ASME Paper No. 2000-GT-0033*

Turney, P. & Halasz, M. (1993). Contextual Normalization Applied to Aircraft Gas Turbine Engine Diagnosis. *Journal of Applied Intelligence*, Vol. 3, pp. 109-129

Volponi, A.J.; DePold, H. & Ganguli, R. (2003). The use of Kalman filter and neural network methodologies in gas turbine performance diagnostics: a comparative study, *Journal of Engineering for Gas Turbines and Power*, Vol. 125, No. 4, pp. 917-924

7

The Recovery of Exhaust Heat from Gas Turbines

Roberto Carapellucci and Lorena Giordano
Department of Mechanical, Energy and Management Engineering
University of L'Aquila
Italy

1. Introduction

Gas turbines have established an important role in the industrial production of mechanical energy owing to the very high power-to-weight ratio achievable with simple-cycle configurations and to the high conversion efficiency that can be obtained in systems that envisage waste heat recovery from the exhaust gases.

Exhaust heat from gas turbines can be recovered externally or internally to the cycle itself [1-4]. Of the various technology options for external heat recovery, the combined gas–steam power plant is by far the most effective and commonly used worldwide. For internal heat recovery, conventional designs are based on thermodynamic regeneration and steam injection, while innovative solutions rely on humid air regeneration and steam reforming of fuel.

In this chapter different techniques for recovering the exhaust heat from gas turbines are discussed, evaluating the influence of the main operating parameters on plant performance.

A unified approach for the analysis of different exhaust heat recovery techniques is proposed. This methodology is based on relationships of general validity, in the context of interest, and on a characteristic plane for exhaust heat recovery, that indicates directly the performance obtainable with different recovery techniques, compared to a baseline non-recovery plant.

Then an innovative scheme for external heat recovery is presented: this envisages repowering existing combined cycle power plants through injection of steam produced by an additional unit consisting of a gas turbine and a heat recovery steam generator.

2. The recovery of exhaust heat

The growing popularity of gas turbines in recent years is attributable to the rapid changes in this technology, which have led to improvements in the design of both the individual components and the system as a whole. These technological advances, that concern important developments in materials, construction techniques, blades cooling, control of pollutant emissions, reliability and availability of machines, have enhanced the performance of simple cycle gas turbines, in terms of electrical efficiency and unit size. They have also led to a significant increase in temperature and flow rates at the gas turbine exit, thus the need for efficient exhaust heat recovery systems.

The waste heat exhausted from gas turbines can be recovered externally or internally to the cycle itself. External heat recovery can be achieved using a bottoming steam power plant (combined cycle). Internal heat recovery involves reusing the thermal energy exiting the turbine, by means of conventional (thermodynamic regeneration and steam injection) or unconventional techniques (humid air regeneration, steam fuel reforming) [5].

In this paragraph the different recovery techniques will be discussed, highlighting for each of them the main thermodynamic and economic features.

2.1 Combined gas-steam cycle power plants

A combined cycle gas turbine (CCGT) is a fossil fuel power plant that combines the Brayton cycle of the gas turbine with the Rankine cycle of the steam turbine. In a typical layout, shown in Figure 1, exhaust heat from the gas turbine, passing through a heat recovery steam generator (HRSG), produces steam that evolves in the bottoming steam cycle. This type of recovery is said to be "direct", because the heat is transferred directly to the working fluid of another system. In order to improve heat recovery in the HRSG, more than one pressure level is generally required. Combined-cycle configurations, with a triple pressure heat recovery steam generator and steam reheat, attain thermal efficiency of more than 55%.

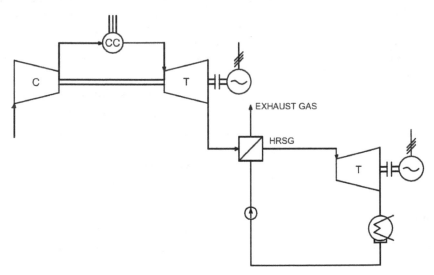

Fig. 1. Schematic diagram of CCGT cycle

In addition to high efficiencies, combined cycle plants have many other advantages, including [6]:

-   low emissions, since natural gas produces no ash or SO_X and smaller quantities of volatile hydrocarbons, CO and NO_X than oil and coal;
-   low capital costs and short construction times (often 2–3 years);

- smaller space requirements than equivalent coal or nuclear power plants;
- flexibility in plant size, ranging from 10 to 750 MWe per combined cycle-unit;
- fast start-up, making it easier to respond to changes in demand.

Recent years have seen a significant increase in the use of combined cycle power plants that, despite the higher fuel cost compared to conventional steam power plants, are currently the best choice in terms of cost per unit of electricity [7].

These plants had already become quite popular in the Italian cogeneration sector. Since the mid 1990s, as shown in Figure 2, installed capacity has grown considerably, bringing the number of sections from 57 (1996) to 131 (2008).

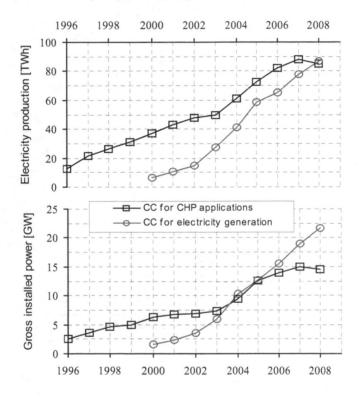

Fig. 2. Combined cycle plants for electric power production and cogeneration in Italy [8]

But since 2000 these plants have found application in the generation of electric power alone, with the installation of 5 sections, increasing to 53 in 2008, with a total installed capacity of around 22000 MW. These systems are characterized by very high efficiency, on average more than 54% [8]. This trend is likely to continue in the future, as shown by the planning permissions granted from 2002 to April 2010, for the construction of new power plants in Italy (Table 1). Indeed, nearly all the plants have efficiencies of above 50%, testifying to the advantages of adopting combined cycle plants.

COMPANY	PLANT LOCATION		MWe	MWt	MWe/MWt
EDISON	MARGHERA		environmental remediation		
	TORVISCOSA		800	1500	53.3%
	ORTA DI ATELLA		780	1340	58.2%
	ALTOMONTE		800	1400	57.1%
	SIMERI CRICHI		800	1360	58.8%
	PIANOPOLI		800	1360	58.8%
	CANDELA		360	650	55.4%
		Total	*4340*	*7610*	*57.0%*
ENIPOWER	FERRERA ERBOGNONE		1040	1850	56.2%
	MANTOVA		780	1370	56.9%
	RAVENNA		785	1370	57.3%
	BRINDISI		1170	2200	53.2%
	FERRARA		800	1400	57.1%
		Total	*4575*	*8190*	*55.9%*
ENEL PRODUZIONE	CASTEL SAN GIOVANNI		80	120	66.7%
	LIVORNO		environmental adaptation		
	CAVRIGLIA/SANTA BARBARA		390	700	55.7%
	CIVITAVECCHIA		fuel change		
EDIPOWER	TURBIGO		repowering		
	PIACENZA		58 MWe for summer peak load		
ENDESA ITALIA	TAVAZZANO		conversion to combined cycle		
	FIUME SANTO		80	220	36.4%
ABRUZZOENERGIA	GISSI		760	1400	54.3%
ACEAELECTRABEL PRODUZIONE	LEINI'		380	700	54.3%
AEM MI -ASM BS	CASSANO D'ADDA		390	700	55.7%
AEM TORINO	MONCALIERI		770	1350	57.0%
ASM BS e AMGS VR	PONTI SUL MINCIO		250	450	55.6%
CALENIA ENERGIA	SPARANISE		800	1400	57.1%
E.ON ITALIA PRODUZIONE	LIVORNO FERRARIS		800	1400	57.1%
ELECTRABEL ITALIA	ROSIGNANO SOLVAY		400	750	53.3%
EN PLUS	SAN SEVERO		390	700	55.7%
ENERGIA	BERTONICO/TURANO LODIGIANO		800	1400	57.1%
ENERGIA MODUGNO	MODUGNO		750	1350	55.6%
ENERGIA MOLISE	TERMOLI		750	1300	57.7%
ENERGY PLUS	SALERNO		780	1370	56.9%
EUROSVILUPPO ELETTRICA	SCANDALE		800	1390	57.6%
IRIDE ENERGIA	TORINO NORD		400	710	56.3%
ITALGEN	VILLA DI SERIO		190	365	52.1%
MIRANT GENERATION PORTOGRUARO	PORTOGRUARO		planning permission expired		
RIZZICONI ENERGIA	RIZZICONI		800	1400	57.1%
SARMATO ENERGIA	SARMATO		47	70	67.1%
SET	TEVEROLA		400	750	53.3%
SORGENIA	APRILIA		750	1350	55.6%
TERMICA CELANO	CELANO		70	100	70.0%
TIRRENO POWER	VADO LIGURE		conversion to combined cycle		
TIRRENO POWER	NAPOLI LEVANTE		400	700	57.1%
VOGHERA ENERGIA	VOGHERA		400	750	53.3%
		Total	*21800*	*38695*	*56.3%*

Table 1. Planning permissions for power plants in Italy, granted from 2002 to April 2010 [9]

2.2 Gas turbine configurations with internal heat recovery

Compared to simple cycle gas turbines, the higher costs of constructing a combined cycle plant are not always offset by higher efficiency, especially for small and medium size plants. The need to combine the high efficiency of combined cycles with the low cost of simple cycles has raised the interest in new technologies that enable internal waste heat recovery from the gas turbine.

Internal heat can be recovered through the working fluid (fuel, air) or an auxiliary fluid (usually water). In the first case internal heat recovery is defined as "direct", in the second as "indirect"[5].

Thermodynamic regeneration is a direct internal recovery technique, since thermal energy is transferred directly from exhaust gas to air at the compressor exit. This produces an efficiency gain due to the reduction in primary thermal energy requirements without changing, as a first approximation, the mechanical power output.

Steam injection on the other hand is an indirect internal recovery technique. In this case the recovered thermal energy is transferred to an auxiliary fluid (water), which is then injected into the combustor. This increases the primary thermal energy required to keep the temperature at the turbine inlet constant, but results in a power increase and, consequently, an efficiency gain.

Direct and indirect recovery can also be combined, as for instance in HAT and CRGT cycles. In humid air plants (HAT) , the saturation of air at compressor exit extends the regeneration margins, thanks to the greater temperature difference between the exhaust gas at turbine exit and the compressed air at regenerator inlet. In chemically recuperated plants (CRGT), exhaust heat is recovered through an endothermic steam-reforming process of the primary fuel. More specifically, a portion of the recovered heat is transferred directly to the fuel, while the remainder is used to produce the required steam.

2.2.1 Heat recovery without auxiliary fluid

In thermodynamic regeneration, the exhaust heat at the turbine exit is used to preheat the air entering the combustion chamber. The heat exchange between the two gas streams is achieved by means of a countercurrent heat exchanger, known as a regenerator or recuperator. Figure 3 shows a schematic diagram of the regenerative cycle.

The thermal efficiency of the Brayton cycle is enhanced since regeneration decreases the heat input required to produce the same net work output. Heat recovery through a gas-to-gas heat exchanger is limited by a characteristic value of the compression ratio, beyond which the temperature of the exhaust gas falls below that of the air at the compressor outlet, thereby deteriorating efficiency.

The efficiency gain achieved through regeneration strongly depends on the heat exchanger effectiveness, defined as the ratio of the actual heat transfer rate to the air and the maximum possible heat transfer rate, that would exist were the heat exchanger to have infinite heat transfer surface area. More specifically, gas turbine efficiency increases with heat exchanger effectiveness, as the air at the combustion chamber inlet is preheated at higher temperatures, resulting in greater fuel savings.

However, to achieve greater effectiveness requires a larger heat transfer area. This translates into higher capital costs and larger pressure drops on both air and gas sides of the heat exchanger, which reduce the turbine pressure ratio and therefore the turbine work.

Generally, the air pressure drop on the high-pressure side should be kept below 2% of the total compressor discharge pressure. The effectiveness of most regenerators used in practice is below 0.85.

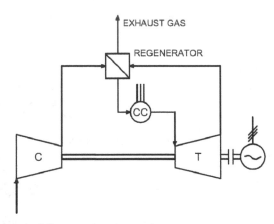

Fig. 3. Schematic diagram of the regenerative cycle

The regenerator takes up a very large area because of the low heat transfer coefficients of air and exhaust gases. This results in high costs and loss of compactness, affecting the main features of gas turbines. In addition, the exhaust gas flow in the regenerator leads to the formation of carbon deposits, resulting in a reduction of heat transfer coefficient, which becomes more pronounced over time.

The regenerator has met with little success in the industrial gas turbines sector, because of the low efficiency gain accompanied by the significantly higher capital costs.

Only in recent years this option has started to make a comeback, as it can be advantageously integrated with other technologies, such as steam injection.

Moreover regeneration is successfully applied in micro - gas turbines, where the low compression ratios are related to the simplicity of turbomachines.

2.2.2 Heat recovery through steam generation

In the steam injection cycle (STIG), proposed by Cheng in 1978 [10], the gases exhausting the turbine are used to produce steam in a heat recovery steam generator, that is then injected into the combustion chamber (Figure 4).

In addition to drastically reducing the formation of nitrogen oxides, steam injection increases both efficiency and power output [11]. The efficiency gain is about 10%, lower than that obtained with a conventional combined cycle, because the steam expands in a less efficient manner in the gas turbine than in the steam turbine. On the other hand, the power increase varies between 50 and 70% [12].

The steam pressure should be sufficient to enable injection into the combustor. Typical values of this parameter range from 1.25 to 1.4 times the maximum pressure of the cycle. In addition, the water used for steam production must be demineralized to minimise salts and oxides content so as to prevent fouling the turbomachinery or chemical attack at high temperatures.

A practical concern with steam injection is water consumption, that typically ranges from 1.1 to 1.6 kg of high purity water per kWh of electrical output. The water purification system required for large scale plant would represent about 5% of total capital costs, whereas running costs add about 5% to the fuel cost [12].

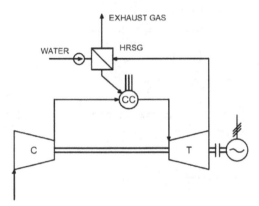

Fig. 4. Schematic diagram of STIG cycle

As the temperature of injected steam must be raised to combustion chamber temperature, a small temperature difference at the approach point (i.e. at boiler outlet) can be advantageous. However, since the evaporation process takes place at constant temperature, high steam temperatures are necessarily accompanied by low heat recovery.

Since the high steam temperature requirements conflict with large heat recoveries, there is a limit to the maximum attainable efficiency with STIG cycles.

Efficiency can be improved using multi-pressure systems whereby the water vapour temperature profile matches that of the exhaust gases more closely, resulting in a better approximation to a reversible process [1]. However, this complicates the design, while plant simplicity is one of the strengths of this technology.

As shown in Figure 5, humid air turbines (HAT) combine regeneration and steam injection. Compared to a traditional regenerative gas turbine, HAT cycle requires the addition of a surface heat exchanger and a saturator. The air at the compressor outlet passes through the heat exchanger, where liquid water is preheated; then air passes through the saturator, where it mixes with the steam, undergoing at the same time a reduction in temperature, as water evaporation absorbs latent heat from the gas stream. The saturation of air at compressor exit extends the regeneration margins, thanks to the greater temperature difference between the exhaust gas exiting the turbine and the compressed air at the regenerator inlet. Moreover humidification reduces the heat capacity difference between air and exhaust gas, resulting in increased efficiency of the regenerator heat recovery [2].

HAT cycles overcome the intrinsic limits of steam injection, further enhancing efficiency.

The main issue of humid air turbines is the difficulty in containing the pressure drops related to the compressed moist air flow through the saturator and the regenerator. The saturator can operate with any clean and filtered water source, as long as the dissolved substances at the water outlet remain below their precipitation concentration under operating conditions [13]. Water consumption is a problem as for steam injection cycles but the consumption rate is only about one third [1].

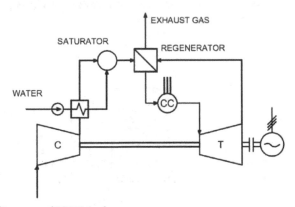

Fig. 5. Schematic diagram of HAT cycle

The chemically recuperated gas turbine (CRGT) is an extension of the steam-injected gas turbine concept [14]. As shown in Fig. 6, the thermal energy available in the exhaust gases is used to promote an endothermic reaction in the primary fuel, that can occur with or without water addition. In the first case the process is called steam reforming, whereas in the second simple decomposition. The process requires the presence of a nickel based catalyst and results in the production of a reformed fuel, composed of CO, CO_2, H_2, excess steam and unconverted fuel, which is fed directly to the combustion chamber.

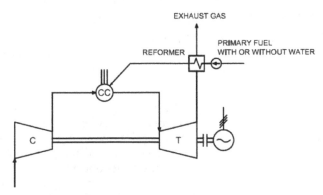

Fig. 6. Schematic diagram of CRGT cycle

The reformed fuel absorbs heat thermally and chemically, resulting in a potentially greater recovery of waste heat than conventional recovery techniques [14]. In fact the thermal energy recovered at a relatively low temperature is made available at a higher temperature, during oxidation of the reformed fuel.

Moreover CRGT cycles have ultra low NOx emissions, due to the large amount of steam contained in the reformed fuel, lowering the temperature in the primary zone of the combustor.

Recently there has been a revived interest in this technology, in view of the potential use of strategic fuels, such as methanol.

3. A unified thermodynamic approach

The different waste heat recovery techniques can be analysed using a unified thermodynamic approach. Heat recovery influences gas turbine performance by means of two concurrent effects. The first is related to the variation of primary thermal energy (ΔQ_1) supplied from the outside, the second to the increase in power (ΔP) produced by the auxiliary fluid, where present.

3.1 Equations for exhaust heat recovery

Denoting with η overall plant efficiency, we can write

$$\frac{\eta}{\eta^*} = \frac{1 + \Delta P / P^*}{1 + \Delta Q_1 / Q_1^*} \tag{1}$$

where η^*, P^* and Q_1^* refer to the non-recovery baseline plant.

Neglecting water pumping power and pressure losses of heat recovery devices, the term ΔP is only related to the auxiliary fluid expanding through the gas turbine. The term ΔQ_1 generally comprises two contributions

$$\Delta Q_1 = \Delta Q_{1,AF} - \Delta Q_{1,DR} \tag{2}$$

The first term $\Delta Q_{1,AF}$ is the additional primary thermal energy required to attain the maximum cycle temperature when an auxiliary fluid is introduced. The second term $\Delta Q_{1,DR}$ refers to the heat reintroduced into the cycle, on the air or fuel side, through direct recovery.

Equation (1) can be taken as a basic relation for characterizing the capabilities of internal and external heat recovery of gas turbines.

For direct external heat recovery, i.e. in the case of combined gas-steam cycles, the term ΔP refers to the steam turbine power output and ΔQ_1 is positive only when supplementary firing is performed.

For direct internal heat recovery, such as thermodynamic regeneration or dry chemical recovery, the power remains practically unchanged ($\Delta P = 0$) and the term $\Delta Q_{1,AF}$ is nil. Therefore the efficiency increase is only due to the reduction in the primary thermal energy supplied ($\Delta Q_1 = -\Delta Q_{1,DR} < 0$). For thermodynamic regeneration $\Delta Q_{1,DR}$ can be immediately interpreted; for dry chemical recovery it represents the energy required by the endothermic decomposition process, recovered from the exhaust gases and transferred to the reformed fuel, by increasing the heating value.

For indirect internal heat recovery, such as steam injection, the introduction of an auxiliary fluid increases power output ($\Delta P > 0$) and primary thermal energy ($\Delta Q_1 = \Delta Q_{1,AF} > 0$). Efficiency increases only if

$$\eta_{FA} = \frac{\Delta P}{\Delta Q_{1,AF}} > \eta^* \tag{3}$$

where the term on the left hand side can be interpreted as the marginal efficiency of the auxiliary fluid. State of the art gas turbine technology satisfies this condition for steam injection, even for low degrees of superheat, but not for water injection, that produces a power increase with an efficiency penalty.

In the case of combined (direct-indirect) internal heat recovery, such as humid air regeneration or steam reforming, efficiency increases as a result of two effects. The first refers to the power output increase (ΔP), the second to the primary energy variation (ΔQ_1), that can be negative or positive in accordance with Eq. (2). Since ΔP and $\Delta Q_{1,AF}$ are proportional to the mass flow rate of the auxiliary fluid, for a given value thereof, the efficiency gains are greater the more $\Delta Q_{1,DR}$ increases.

3.2 A performance plane for exhaust heat recovery

Using Eq. (1) it is possible to define a characteristic plane, that allows to compare different techniques for recovering exhaust heat from gas turbines, highlighting their application limits.

As shown in Figure 7, the performance plane of waste heat recovery indicates the trend of the ratio η/η^* as a function of $\chi = \Delta Q_1/Q_1^*$ and $\pi = \Delta P/P^*$. The relation between η/η^*, χ and π does not depend on gas turbine characteristics, which are instead introduced by two other families of curves. These define the conditions for constant values of the direct recovery parameter, defined as $\xi = \Delta Q_{1,DR}/Q_1^*$, and for those of the non-dimensional flue gas temperature, defined as $\tau = T_{FG}/T_{FG}^*$ [5]. From Eqs. (2) and (3), we can derive a relationship among different non-dimensional parameters of internal heat recovery

$$\chi = \pi\left(\eta^*/\eta_{AF}\right) - \xi \tag{4}$$

This establishes, for a given gas turbine (η^*) and recovery technique (ξ and η_{AF}), the relationship between χ and π.

For $\pi = 0$ – simple direct recovery – from Eq. (4) we get $\chi = -\xi$; therefore, each point on $\pi = 0$ curve of Fig. 11 is characterized by a different ξ value.

Each point P on this curve defines an envelope of curves at constant ξ, but characterized by different η_{AF} values. Combining Eqs. (1) and (4), we get

$$\left.\frac{\partial(\eta/\eta^*)}{\partial\chi}\right|_{\xi=\text{cost}} = \frac{1}{(1+\chi)^2}\left[\frac{\eta_{AF}(1-\xi)}{\eta^*} - 1\right] \tag{5}$$

Equation (5) defines the slope of curves with constant ξ, at each value of η_{AF} related to the thermodynamic conditions of auxiliary fluid at the combustor inlet.

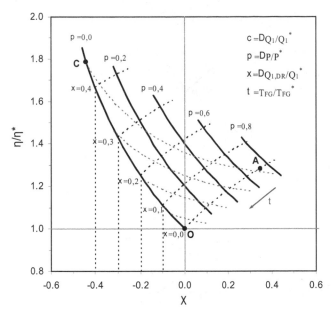

Fig. 7. A performance plane for exhaust heat recovery

The curve with the maximum slope (i.e. maximum η_{AF} value) is obtained at the maximum auxiliary fluid temperature, that occurs, in the case of steam injection, inside the combustor at the maximum degree of superheat permitted by the steam generator, as well as for steam reforming at the maximum temperature allowable by the exhaust gas at the turbine exit.

Decreasing η_{AF}, that is for a lower enthalpy of the auxiliary fluid introduced into the combustor, the slope of curves at constant ξ diminishes eventually becoming negative in the case of water injection.

Figure 8(a), for instance, shows two η_{AF} curves for steam injected at the maximum degree of superheat ($\eta_{AF,1}$) and under saturated conditions ($\eta_{AF,2}$). For a given plant with no heat recovery (P^*, η^*), a generic point Q on the characteristic plane, as shown in Figure 8(b), may represent different internal heat recovery techniques characterized by various combinations of direct and indirect recovery ($\xi_1 < \xi_2$ and $\eta_{AF,1} > \eta_{AF,2}$), based on plant configurations defined by different value of ξ and η_{AF}.

The constant τ curves indicate the extent of the recovery. For a fixed value of τ, the maximum efficiency increase is obtained for $\pi=0$. Instead, when an auxiliary fluid is introduced, at constant τ, the efficiency increase is lower due to the unrecoverable latent heat of steam at the turbine exit.

Assuming a limit value for the flue gas temperature, the corresponding curve, together with the curves at $\pi=0$ and $\xi=0$ define a characteristic region (OAC in Fig. 7) which represents the possible recovery conditions. Each point inside this region does not represent a specific plant configuration, since the same performance can be obtained with different heat recovery techniques.

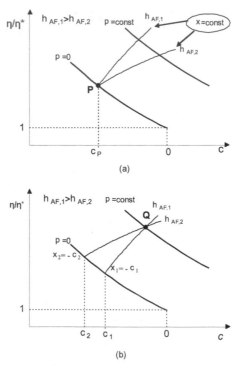

Fig. 8. Influence of marginal auxiliary fluid efficiency on combined recovery capabilities

Points on the curve $\pi=0$ (curve OC in Fig. 7) therefore indicate plant configurations with direct recovery $(\chi=-\xi)$ alone, those on the curve OA $(\xi=0)$ on the contrary, solutions with indirect recovery $(\chi=\pi\ \eta^*/\eta_{AF})$ alone and are characterized by positive values of χ denoting an increase of primary energy supplied to the cycle.

The plane region above curves OC and OA represents conditions for combined direct and indirect recovery. Negative values of χ, denoting a net reduction of primary energy to the cycle, are possible if direct recovery effects predominate over those associated with steam injection. One limitation to the extension of the OAC region is the minimum flue gas temperature attainable inside the stack, also taking into account the characteristics of the single internal heat recovery techniques (regenerator effectiveness, saturation conditions, pinch point at HRSG, steam-to-methane ratio at reformer). A further restriction may also arise from operational problems with existing combustor and turbomachinery, especially if high steam flow rates are injected [5].

4. Capabilities of exhaust heat recovery techniques

The characteristic plane of heat recovery can be used to determine the capabilities of different internal and external heat recovery techniques for a variety of gas turbines.

Having defined the non-recovery gas turbine, the construction of this plane requires the evaluation of performance increases achieved by different heat recovery configurations compared to the baseline gas turbine. For this purpose the General Electric software GateCycle has been used [15]. Using this modelling tool specific plant configurations have been developed to simulate the non-recovery baseline simple cycle (SC), the regenerative cycle (RG), the steam injected cycle (SI), the regenerative steam injected cycle (RG+SI), the humid air regenerative cycle (HAT) and the chemically recuperated cycle (CRGT). In all cases with exhaust heat recovery, combustor and turbomachinery design data are taken to be the same as for the corresponding simple non-recovery cycle, while operating data for each heat recovery device have been examined over significant ranges.

In order to evaluate the influence of pressure ratio and turbine inlet temperature on capabilities of different recovery techniques, the characteristic plane of heat recovery has been defined with reference to four non-recovery gas turbines, that differ in terms of pressure ratios and turbine inlet temperatures.

Referring to these characteristic planes, represented in Figure 9, the different internal and external heat recovery techniques will be discussed in more detail in the following subsections, highlighting the influence of pressure ratio and temperature inlet temperature on efficiency increase.

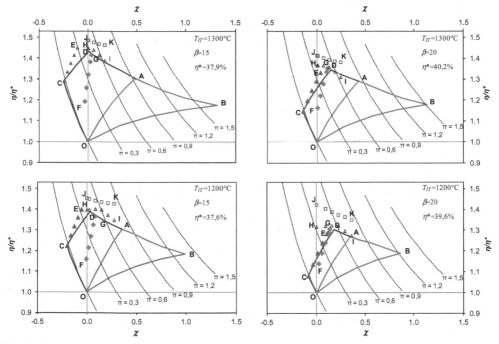

Fig. 9. Characteristic plane of heat recovery for different non-recovery gas turbines (Internal heat recovery - *OC : Thermodynamic regeneration, OAB: Steam injection, CD: Regeneration with steam injection, CE: Humid air regeneration (HAT), FG: Steam methane reforming.* External heat recovery - *HI: Combined cycle with one pressure level HRSG, HI: Combined cycle with three pressure levels HRSG and reheat.*)

4.1 Direct external heat recovery

Curves HI and JK represent the performance obtained for external heat recovery through a bottoming steam cycle (combined cycle). In particular, the curve HI refers to the case of a combined cycle with one pressure level at the HRSG, whereas curve JK to a combined cycle with three pressure levels and reheat. Points H and J correspond to the case of combined cycle without supplementary firing, the primary energy supplied from outside ($\chi = 0$) remaining the same. Points I and K correspond instead to the case of supplementary firing using exhaust gases exiting the turbine and setting the limit temperature at the duct burner exit to 800 ° C.

Without supplementary firing, the combined cycle with three pressure levels HRSG and reheat (point J) is the leading solution in power generation, providing efficiencies of 54.6%-56.8% in the explored range of β and T_{IT}, higher by 2-3 percentage points than those obtained with one pressure level HRSG (point H). With supplementary firing, the further power increase is obtained at the expense of efficiency (I and K). Only in the case of one pressure level HRSG, with modest additional combustion and low values of T_{IT}, a slight improvement in performance is achieved, as a result of the greater recovery feasible at low temperatures.

4.2 Direct internal heat recovery

Direct internal heat recovery, achieved by transferring thermal energy from the gas turbine exhaust to the compressed air upstream from the combustor, is represented by curve OC on the characteristic plane.

At point O the temperature difference on the hot side of the regenerator (ΔT_{RG}) is zero. Since there is no heat exchange between exhaust gas and air at the compressor exit, gas turbine efficiency coincides with the baseline simple cycle with no heat recovery. Moving along the curve OC the temperature difference increases, as well as regenerator effectiveness and gas turbine efficiency, reaching a maximum value at point C, corresponding to a temperature difference of 40 ° C.

Neglecting pressure losses, point C shows a slight power derating with respect to the reference plant (point O) due to the smaller amount of primary fuel introduced into the combustor.

To evaluate the influence of pressure ratio and the turbine inlet temperature on efficiency increase through thermodynamic regeneration, it suffices to compare the OC curve of the different baseline non-recovery gas turbines. As shown in Figure 9, efficiency gain is maximised at low pressure ratios and high turbine inlet temperatures. Focusing the attention on point C, we see that direct recovery parameter ($\xi = -\chi$) increases from 0.08 at low maximum gas temperature (T_{TI}=1200°C) and high pressure ratio (β=20) up to 0.24 at T_{TI}=1300°C and β=15. Consequently, the efficiency ratio η/η^* passes from 1.08 to 1.29, the efficiency η of the regenerative cycle from 42.9% to 48.9% and the flue gas temperature T_{FG} from 470°C to 420°C.

4.3 Indirect internal heat recovery

The points inside the region OAB (Fig. 9) represent performance achievable with waste heat recovery performed by steam injection upstream from the combustion chamber. This region is bounded by two curves at $\xi=0$ (OA and OB) and by the curve AB corresponding to the minimum temperature difference at pinch point of the heat recovery steam generator ($\Delta T_{PP}=10°C$). As mentioned above, the slope of the curve with $\xi=0$ depends on the gas turbine plant with no heat recovery, η^* and on marginal efficiency η_{AF}.

In particular the curve OA refers to the case of maximum marginal efficiency η_{AF} obtained by injecting steam superheated to the same temperature as the turbine exhaust, while curve OB refers to the case of minimum η_{AF}, corresponding to injection of saturated steam. The entire region can be covered varying the hot side temperature difference in the superheater and the minimum temperature difference in the evaporator.

Efficiency gains due to steam injection diminish with steam temperature, while both steam mass flow rate and power produced increase. In practice the power increase is limited by problems associated with the large water requirements and compressor-turbine matching.

To evaluate the influence of β and T_{TI} on steam injection capabilities, points A and B are examined. In both cases, steam mass flow rate as well as power increase with temperature, while the opposite trend is observed when pressure ratio is increased.

In particular, at point A (superheated steam) the mass flow rate increase passes from a minimum of 20% at low maximum gas temperature ($T_{TI}=1200°C$) and high pressure ratio ($\beta=20$) to 30% at $T_{TI}=1300°C$ and $\beta=15$. On the other hand, at point B (saturated steam) the mass flow rate increase ranges from 34 to 55%, under the same pressure and turbine inlet temperature conditions.

4.4 Combined internal heat recovery

Performances associated with points inside the region OCDA (Fig. 9) can only be obtained considering recovery techniques that combine direct (thermodynamic regeneration) and indirect recovery (steam injection). The effects associated with the auxiliary fluid occur in different ways, with regard to marginal   efficiency value (η_{AF}), which depends on the thermodynamic conditions of the auxiliary fluid upstream from the combustion chamber.

The region OCDA can be covered varying the hot side temperature difference in the regenerator and the minimum temperature difference in the evaporator.

Heat can also be indirectly recovered using unconventional techniques, such as humid air regeneration and steam reforming of the fuel.

The performance of HAT plants is represented by the points on curve CE. Keeping the hot side temperature difference in the regenerator at 40°C, curve CE has been obtained increasing the mass of water introduced into the saturator from zero (point C) to the maximum permissible value for saturation of the compressed air upstream from the regenerator (point E). On this curve the value of ξ, defined at point C, remains constant, since this parameter is established by the capabilities of the plant with no heat recovery with respect to thermodynamic regeneration.

Gas turbine plants with chemical recovery are represented on curve FG, where methane is used as primary fuel. The methane reforming process is described by the following two reactions:

$$CH_4 + H_2O \rightarrow CO + 3H_2 \tag{6}$$

$$CO + H_2O \rightarrow H_2 + CO_2 \tag{7}$$

The reforming reaction (Eq. (6)) is highly endothermic and is favoured by higher temperatures, lower pressures and higher steam-to-methane mole ratios. The water gas shift reaction (Eq. (7)), slightly exothermic, favours lower temperatures and is unaffected by pressure [16].

Keeping the hot side temperature difference in the reformer at 40°C, curve FG refers to steam reforming performed varying the steam-to-methane mole ratio from the stoichiometric value n=2 (point F) to the maximum value corresponding to the minimum pinch point temperature difference (point G).

Curve FG is characterized by a steep slope, for increasing values of ξ. In fact, the thermal energy directly recovered, denoted with ξ, is the chemical energy variation associated with the reforming process that increases with steam-to-methane mole ratio (n).

The influence of β and T_{TI} on combined internal heat recovery capabilities are discussed focusing the attention on points representing maximum heat recovery conditions for each solution: point D for regenerative steam injected cycle, point E for humid air regenerative cycle, point G for chemically recuperated cycle.

For regenerative steam injected cycle (point D) at high turbine inlet temperature (T_{TI}=1300°C) and low pressure ratio (β=15), direct and indirect recovery have comparable effects on χ; therefore primary fuel energy introduced into the cycle remains practically unchanged, while efficiency exceeds 54.3%, due to the significant power increase (π=0.45). On the contrary, at T_{TI}=1200°C and β=20, effects of indirect recovery prevail, producing a higher power increase (π=0.54) and a lower efficiency gain (η/η^*=1.3).

For humid air regenerative cycle (point E), efficiency gains achievable are higher than the regenerative steam injected cycle (point D), particularly at high maximum gas temperature and low pressure ratio. At T_{TI}=1300°C and β=15, HAT plants attain efficiencies of up to 55%, limiting the relative mass flow rate increase in the turbine to 10%.

In the case of the chemically recuperated cycle (point G), since steam methane reforming reactions prefer low pressure and high temperature, the greatest efficiency gains are obtained at high turbine inlet temperature and low pressure ratio. At T_{TI}=1300°C and β=15, methane conversion ratio is close to 55% and, consequently, efficiency increases up to 53.8%.

4.5 Limits of indirect internal heat recovery

Steam injection in a gas turbine is affected by operational constraints related to compressor-turbine matching, defined on the basis of the characteristic curves of turbomachines. Figure 10 shows a typical axial compressor map, bounded above by the surge line and below by the choke line. Operation limits in the surge region are due to an increase in the angle of incidence between the fluid and the compressor blades, produced by a decrease in fluid flow rate or an increase in rotational speed. Any excessive increase of the angle of incidence may cause fluid separation and flow reversal, generally accompanied by strong noise and violent vibrations which can severely damage the machinery. In order to avoid this instability phenomenon, the axial compressor operates in nominal conditions, with a certain margin from the surge region.

However, when steam injection occurs, both the mass flow rate and the turbine inlet pressure increases. The compressor follows the turbine behaviour by increasing the compression ratio and, consequently, approaching the surge line.

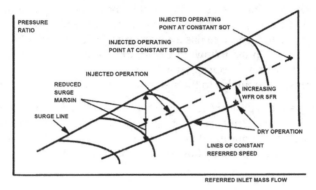

Fig. 10. Gas turbine operating line in dry and wet conditions [17]

Therefore the injection rate must be regulated to keep the pressure ratio below the surge line. In practice, for existing gas turbines, injected steam flow rate is limited to 10% of the compressor inlet air flow [18].

Moreover, the water introduced into the gas turbine may create problems associated with availability and treatment and with the mass flow rate increase through the expander.

In the presence of low-temperature heat users, the exhaust gases could be cooled down to 50 ° C, in order to achieve partial water condensation.

The water can be condensed in an indirect surface heat exchanger, that uses water or ambient air to cool the exhaust gas or in a direct-contact condenser, where water is sprayed into the exhaust gas [2].

Large amounts of water are required for partially condensing steam, so these power plants should be located near to water sources (sea, lakes, rivers). In cases of low water availability, "closed loop" refrigeration is conducted, sending the water at the condenser outlet to a cooling tower.

4.5.1 Effect of limits on the characteristic plane of heat recovery

To ensure proper operation of the gas turbine, the limit on maximum flow rate increase at the turbine inlet, results in a reduction of the maximum performance achievable by the different internal heat recovery techniques.

In this regard, Figure 11 shows the characteristic plane of heat recovery for a baseline gas turbine T_{IT} = 1300°C and $\beta=15$. In this plane, the dash-dot line curve refers to an increase in mass flow rate at the turbine inlet of 10%. As shown in Figure 11, the performance region for steam injection is significantly reduced, passing from OAB to OA'B'; similarly the region related to combined recovery (obtained by means of thermodynamic regeneration and steam injection) is reduced from OCDA to OCD'A'.

Indirect external heat recovery

The issue of limiting the maximum flow rate increase at the turbine inlet can be overcome by carrying out indirect external heat recovery, represented by curve OL in Figure 11. This recovery option, which will be discussed in detail in the following paragraph, is achieved by injecting superheated steam produced in the combustion chamber of an existing combined system. The existing combined cycle has a three pressure levels HRSG, with characteristics similar to those indicated in the following paragraph. Efficiencies defined by curve OL are assessed in marginal terms, i.e. appropriately taking into account only primary energy and power output increases attributable to the steam injected into the combustion chamber.

The curve OL shows the typical trend of internal recovery through steam injection (curve OA'). However, this recovery option is not affected by limits on the maximum steam flow rate, the gas turbine being appropriately sized for integration with the existing combined cycle, in order to keep steam injection flow rates below 10% of the air flow at the compressor inlet.

Moreover, it is interesting to note that the slope of curve OL is greater than curve OA, due to the improved performance of the combined cycle compared to the simple gas turbine. This allows to better exploit the injected steam, leading to higher efficiency gains.

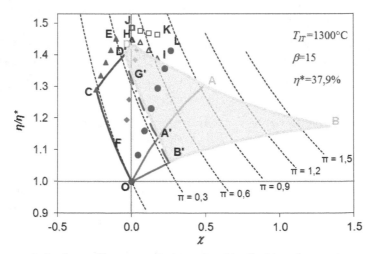

Fig. 11. Characteristic plane of heat recovery restricted by limiting the maximum injected steam flow rate to 10% of the compressor inlet air flow
(*OL: Steam injection within an existing combined cycle*)

5. Repowering of combined cycle power plants through steam injection

The analysis carried out by means of the characteristic plane of heat recovery has shown that steam injection cannot compete with combined heat recovery techniques, such as humid air regeneration or fuel steam reforming.

However, steam injection can be seen as the only indirect recovery technique introducing an innovative scheme, whereby steam injection is not used as a means of traditional internal heat recovery, reintroducing steam into the gas turbine combustion chamber. On the contrary heat is recovered externally, generating steam for repowering an existing combined cycle.

5.1 Description of the repowering scheme

The design concept of the proposed scheme is shown in Figure 12. As discussed in [19], it is based on the addition of a gas turbine and a heat recovery steam generator to an existing combined cycle. The integration of new components into the baseline combined cycle is achieved by injecting the steam generated by the additional HRSG into the combustion chamber of the existing combined cycle.

The power increase is, thus, the sum of the power produced by the new gas turbine and the additional power generated in the original combined cycle, by the additional steam flow in the gas turbine and steam cycle.

The pressure required to inject steam into the turbine is relatively low compared to that usually employed in steam turbines. Therefore the additional heat recovery steam generator can have a single pressure level and a low pinch point, thereby reducing stack temperature (close to 120°C) and hence increasing exhaust heat recovery.

Another significant benefit of this solution is the ability to generate additional power without the need to find new sites, simply improving utilization of electricity generation sites where combined cycle plants are already installed, without affecting their excellent performance and environmental compatibility. Another significant feature of the proposed repowering scheme is its operational flexibility. Because of the inherent flexibility of the gas turbine, the entire additional section can be switched off in a short time, yielding a part load efficiency equal to that of the original plant. An international (PCT) patent application has been filed for the proposed repowering scheme [20].

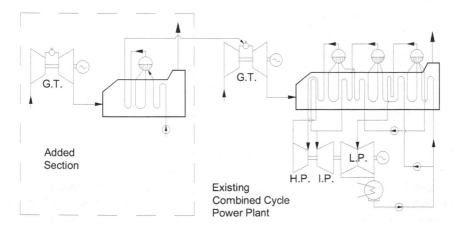

Fig. 12. Schematic diagram of combined cycle repowering through steam injection

The major drawback of this configuration is water consumption related to the flow entering the new HRSG, which is inevitably lost at the stack. This can limit the applicability of the present scheme to sites with large fresh water availability, though the specific water requirements are fairly low, as shown in [19].

Moreover, if a low temperature thermal load is available near the power plant, water can be recovered through steam condensation. The implementation of a water recovery technique has to be carefully evaluated, because of the very large size and the very low temperature level of such a heat sink.

5.2 Discussion of repowering performance

The scheme described above has been numerically studied using a modular simulation code [19], based on fundamental thermodynamic relations. In this simulation code each power plant component (gas and steam turbines, condensers, heat exchangers etc.) is modelled by means of mass and energy balances. A detailed analysis has thus been carried out, simulating the proposed repowering scheme with real data from present day power generating plants [19].

Several gas turbine models, both heavy-duty and aeroderivative units, were tested to assess the feasibility of repowering, the power augmentation achievable and its influence on energy conversion efficiency. The results of the analysis have shown that a power increase of up to 50%, with respect to the existing combined cycle plant, can be achieved. Moreover the additional electricity obtained from repowering is generated at high efficiency (49-52%), though the added section is not so efficient, while the cost of electricity is comparable with that of existing combined cycles. The analysis also showed that the proposed repowering scheme offers a variety of power control strategies and, hence, the possibility of achieving good part load behaviour, especially with the addition of a number of small aeroderivative gas turbines.

A further analysis has been carried out to evaluate the influence of the main added turbine features (β and T_{TI}), added HRSG operation (steam degree of superheat) and steam-to-air mass ratio (μ) on repowering performance.

For this purpose a repowering unit (GT and HRSG) has been added to a given baseline natural gas combined cycle plant, designated GE S109FA. It consists of a single General Electric gas turbine type PG9351FA, with design performance summarized in Table 2. Added gas turbine data have been taken to represent General Electric F-series gas turbines, while added HRSG always produces the maximum amount of steam, 10°C being the minimum temperature difference at pinch point and 90°C the minimum gas temperature within the stack.

Repowering performance has been evaluated in terms of incremental variables, i.e. marginal power output P and marginal efficiency η. The first is the power of the added gas turbine plus the power increase in the existing combined cycle due to steam injection (including both GT and ST contributions), the second is the ratio between the marginal power output and the marginal primary fuel power related to added GT and to steam injection into existing CC.

CC model designation	GE S109FA
Number and model of GT	1 x PG9351FA
Gas turbine	
Pressure ratio	15.4
Exhaust gas temperature, °C	610.6
Exhaust gas mass flow rate, kg/s	626.6
Net output, MW	254.2
Net efficiency, %	37.1
Steam cycle	
HP steam pressure, bar	125.1
HP steam mass flow rate, kg/s	69.7
IP steam pressure, bar	28.0
IP steam mass flow rate, kg/s	16.1
LP steam pressure, bar	4.2
LP steam mass flow rate, kg/s	9.8
Condenser pressure, kPa	5.1
Net CC output, MW	386.7
Net CC efficiency, %	56.3

Table 2. Design performance of baseline combined cycle

Varying β (from 10 to 30), T_{TI} (from 1200 to 1600°C) and degree of superheat (saturated and superheated steam), the repowering unit has always been rated such that the amount of steam generated for injection into the existing combined cycle matches the required μ value. To avoid compressor and turbine matching problems, considering that General Electric has offered injection for power augmentation for 40 years on all of its production machines [18], a steam-to-air mass ratio μ=5% is assumed, corresponding to 30.6 kg/s of steam injected into the GE S109FA combustor.

As shown in Figure 13, injection of superheated steam produces higher values of marginal power and efficiency, for any β and T_{TI}.

Marginal power output increases with β, while it is little influenced by T_{TI}; for a pressure ratio of 30, superheated steam injection produces a power increase of 165 MW, of which about 60% (102 MW) produced by the added gas turbine. On the contrary, marginal efficiency is strongly influenced by T_{TI}, especially at high pressure ratios. In this regard, for the added gas turbine operating at β=30, marginal efficiency attains 53.1% at T_{TI}=1200°C and 57.4% at T_{TI}=1600°C.

More interestingly, the proposed repowering scheme offers the possibility of maintaining high efficiency over a wide range of marginal power outputs. In fact, marginal efficiency is strongly influenced by existing CC and added GT characteristics, but only slightly by the steam-to-air mass ratio μ. As shown in Figure 14, by varying μ from 3% to 9%, repowering can generate a marginal power output of up to 100 MW and 300 MW, respectively.

Therefore, though the steam mass flow rate for injection is limited by compressor and turbine matching problems or water availability and treatment requirements, the proposed repowering scheme could be beneficially implemented, as it is still characterized by high marginal efficiency and significant marginal power.

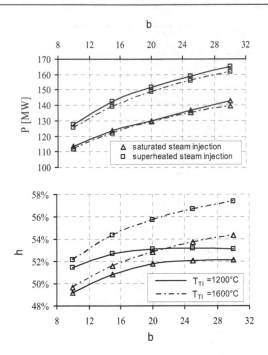

Fig. 13. Performance of CC repowering scheme in terms of marginal efficiency and power

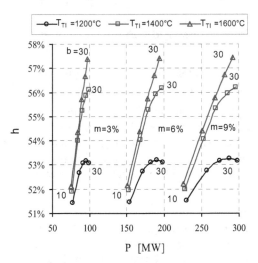

Fig. 14. Influence of steam-to-air mass ratio on performance of CC repowering scheme

Referring again to the combined cycle GE S109FA, Table 3 summarizes the improved performance that can be achieved adopting the repowering scheme for specific commercial gas turbines. In particular, two heavy-duty gas turbines (Siemens and Westinghouse V64.3A 401), with mechanical power in the 70-85 MW range, and aeroderivative (GE LM6000) of smaller size are considered. For the latter, the effects of repowering are also assessed for integration with two or three gas turbine units.

The results obtained show that marginal efficiency is kept above 50%, even with small gas turbine units (around 10MW), while the power increase achieved (marginal power) is directly dependent on the flow rate of injected steam, as shown in Fig.15. Additional gas turbine accounts for about 60% of power increase – 56% for heavy duty GT and 64% for aeroderivative GT.

GT Number and model	GE LM6000	2 x GE LM6000	3 x GE LM6000	Siemens V64.3A	Westing. 401
Added GT power, MW	43.4	86.8	130.2	70.0	85.9
Added GT efficiency, %	41.3	41.3	41.3	36.8	36.5
Injected steam, t/h	45.9	91.7	137.6	99.3	119.8
Steam-to-air mass ratio, %	2.1	4.2	6.3	4.5	5.5
Marginal power, MW	68.2	137.0	205.9	124.0	151.0
Marginal efficiency, %	51.2	51.4	51.4	50.7	50.4

Table 3. Marginal efficiency and marginal power of repowered gas cycles

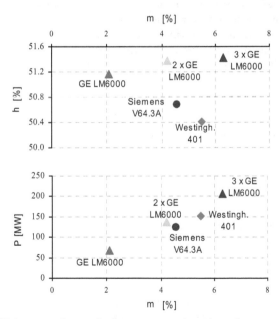

Fig. 15. Marginal efficiency and marginal power as a function of steam-to-air mass ratio

6. Conclusions

Different techniques for the internal or external recovery of exhaust heat from gas turbines have been investigated. Internal heat recovery techniques can be conventional (regeneration, steam injection) or unconventional (humid air regeneration, steam reforming), while external heat recovery can be performed using a steam bottoming cycle (combined cycle).

In order to compare the capabilities of the different solutions, a characteristic plane of exhaust heat recovery, based on a unified analysis approach, has been introduced. The performance plane of exhaust heat recovery is an effective tool for comparing various design solutions that are conceptually different and not directly comparable. On this plane each recovery technique is identified by a region, whose position and extent depends on typical parameters and characteristics of the baseline gas turbine, as well as on limitations related to minimum stack gas temperature and maximum mass flow rate increase in the turbine.

The characteristic plane indicates directly the performance obtainable with various heat recovery techniques. The analysis carried out has shown that performances close to combined cycle plants can only be achieved with combined recovery techniques (humid air regeneration or steam reforming of fuel), where the efficiency penalty is small at high maximum gas temperatures and low compression ratios.

Conventional combined recovery techniques (regeneration and steam injection) can compete with combined cycle plants at low turbine inlet temperature (1200°C), as they offer greater design simplicity, in spite of an efficiency penalty of a few percentage points.

Lastly, from the unified thermodynamic approach an innovative repowering scheme has been proposed. This allows to repower existing combined gas-steam power plants through the addition of a gas turbine and a one-pressure level HRSG, that feeds the output steam to the combustor of an existing gas turbine. This scheme significantly increases power output (50%) with fairly high marginal efficiency, in spite of the relative simplicity of the added components.

7. Nomenclature

n	steam-to-methane mole ratio
P	power
Q_1	primary thermal energy
T	temperature

Greek letters

β	compressor pressure ratio
η	efficiency
μ	steam-to-air mass ratio
ς	direct recovery parameter
π	relative variation of power
τ	non dimensional flue gas temperature
χ	relative variation of primary thermal energy

Subscripts

AF	auxiliary fluid
DR	direct recovery
FG	flue gas
RG	regenerator
TI	turbine inlet

Acronyms

CCGT   combined cycle gas turbine
CRGT   chemically recuperated gas turbine
HAT   humid air regeneration
RG   thermodynamic regeneration
SI   steam injection

8. References

[1] Heppenstall T, Advanced gas turbine cycles for power generation: a critical review, *Appl Therm Eng*, Volume 18, Issue 9-10 (1998) 837-846.

[2] Jonsson M, Yan J, Humified gas turbines – a review of proposed and implemented cycles, *Energy*, Volume 30, Issue 7 (2005) 1013-1078.

[3] Wei H, Hongguang J, Na Z, Xiaosong Z, Cascade utilization of chemical energy of natural gas in an improved CRGT cycle, *Energy*, Volume 32, Issue 4 (2007) 306-313.

[4] Kuroki H, Hatamiya S, Shibata T, Koganezawa T, Kizuka N, Marushima S, Development of elemental technologies for advanced moist air turbine system, *J. Eng. Gas Turbines Power*, Volume 130, Issue 3 (2008).

[5] Carapellucci R, A unified approach to assess performance of different techniques for recovering exhaust heat from gas turbines, *Energy Convers Manag* 50 (2009) 1218-1226.

[6] Poullikkas A. Parametric study for the penetration of combined cycle technologies into Cyprus power system. *Appl Therm Eng* 2004;24:1675-85.

[7] Lozza G, Turbine a gas e cicli combinati, Progetto Leonardo Bologna.

[8] GRTN, Dati statistici sull'energia elettrica in Italia, 2008, Gestore Rete Trasmissione Nazionale, www.grtn.it.

[9] Fonte: Rapporto trimestrale del Ministero dello Sviluppo Economico, http://www.sviluppoeconomico.gov.it

[10] Cheng DY, Regenerative parallel compound dual-fluid heat engine. US Patent 4,128,994; 1978

[11] Macchi E, Consonni S., Lozza G, Chiesa P. An assessment of the thermodynamic performance of mixed gas-steam cycles: Part A – Intercooled and steam-injected cycles. *ASME J Eng Gas-Turbines Power* 1995;117:489-98.

[12] Poullikkas A , An overview of current and future sustainable gas turbine technologies, *Renew and Sustain Energy Reviews* 9 (2005) 409-443.

[13] Traverso A, Massardo AF. Thermodynamic analysis of mixed gas–steam cycles. *Appl Therm Eng* 2002;22: 1-21.

[14] Carcaschi C., Facchini B., Harvey S., Modular approach to analysis of chemically recuperated gas turbine cycles, *Energy Convers Manag* 39 (1998) 1693-1703.

[15] GE Energy, Gate Cycle manual 2005, California, GE Enter Software LLC.

[16] Carapellucci R, Milazzo A, Thermodynamic optimization of a reheat chemically recuperated gas turbine, *Energy Convers Manag* 46 (2005) 2936-2953.

[17] Walsh P.P. , Fletcher P., Gas Turbine Performance, ASME Press, 2004

[18] Johnston JR. Performance and reliability improvements for heavy-duty gas turbines, GER-3571H, GE Power Systems, Schenectady, NY; November 2000.

[19] Carapellucci R, Milazzo A. Repowering combined cycle power plants by a modified STIG configuration. *Energy Convers Managem* 2007;48(5):1590-600.

[20] Carapellucci R, Milazzo A, Scheme of power enhancement for combined cycle plants through steam injection. International Patent (PCT) application n. PCT/IT2006/000332, WO/2006/123388, November 23; 2006.

8

Models for Training
on a Gas Turbine Power Plant

Edgardo J. Roldán-Villasana and Yadira Mendoza-Alegría
Instituto de Investigaciones Eléctricas, Gerencia de Simulación
México

1. Introduction

In México, near 15% of the installed electrical energy of CFE, the Mexican Utility Company is based on gas turbine plants. The economical and performance results of a power plant are related to different strategies like modernisation, management, and the training of their operators.

The Advanced Training Systems and Simulation Department (GSACyS) of the Electrical Research Institute (IIE) in México is a group specialised in training simulators that designs and implements tools and methodologies to support the simulators development, exploitation, and maintenance. The GSACyS has developed diverse works related with the training. The main covered areas by the IIE developments are: computer based training systems, test equipment simulators, and simulators for operators training.

To use real time full scope simulators is one of the most effective and secure way to train power plant operators. According to Hoffman (1995), by using simulators the operators can learn how to operate the power plant more efficiently. In accordance with Fray and Divakaruni (1995), even not full scope simulators are successfully used for operators training.

Some advantages of using simulators for training are the ability to train on malfunctions, transients and accidents; the reduction of risks of plant equipment and personnel; the ability to train personnel on actual (reproduced) plant events; a broader range of personnel can receive effective training, and eventually, high standard individualised instruction or self-training (with simulation devices planned with these capabilities).

In the case of simulators, situations "what would have happened if..." arises when a cost benefit analysis is sought. A classical analysis for fuel power plants simulators (Epri, 1993) identified profit of simulators in four classes: availability savings, thermal performance savings, component life savings, and environmental compliance savings. A payback of about three months was estimated. Most often, the justification for acquiring an operators training simulator is based on estimating the reduction in losses (Hosseinpour and Hajihosseini, 2009). It is not difficult to probe the benefits for high-capacity plants where savings means millions of dollars for a few days of un-productivity. Besides, the ability of the simulator to verify the automation system and provide operators with a better understanding of a new process must be addressed. Using a simulator help operators to improve the skills to bring the plant up and down, thus shortening start-ups significantly and improving the proficiency of less-experienced operators in existing plants.

In 2000 the CFE initiated the exploitation of a Combined Cycle Power Plant Simulator (CCS) developed by the IIE based on ProTRAX, a commercial tool to construct simulators. There is no full access to the source ProTRAX programs and the CFE determined to have a new combined cycle simulator using the open architecture of the IIE products. The new simulator was decided to be constructed in two stages: first the Gas Turbine (GT) part followed by the Heat Recovery Steam Generation (HRSG) part.

In this chapter a summary of the GT simulator development and its modelling characteristics are described. Stochastic and discrete events models are not considered, but deterministic models of industrial processes are contemplated.

2. Prior modelling approaches

A model to simulate a process may be developed using different approaches, basically depending on the use the model will be intended for. Certainly, trying to classify the different ways a model may be designed is a very difficult task. However, a model purpose may be: analysing, designing, optimisation, instruction, or training. The techniques to construct a model may vary, to mention two extreme situations, from governing principles based on differences or finite elements to obtain very detailed physical models, to curves fitting for limited empirical models. The real time modelling approach, particularly for operators training like the model presented in this chapter, lies somewhere in between.

Models for operation training are not frequently published, they belong to companies that provide simulation services and it is proprietary information (see, for example, Vieira *et al.*, 2008). Nevertheless, some models are available in recent publications as described in this section. All the revised works report to have a gas turbine system as presented in Figure 1.

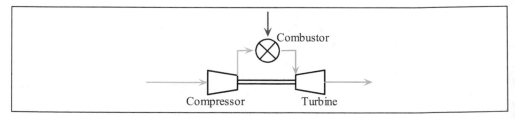

Fig. 1. Common representation of a gas turbine system

A summary of some published models is given in Table 1. None of the works revised here mentioned anything about real time execution.

In the present work, the total plant was simulated including all the auxiliary systems. For the composition of supply gas twenty components were included. There were simulated the variables in the 20 control screens and the tendency graphs. For example, the compressor-turbine system was simulated considering the schematics presented in Figure 2.

Authors (Reference)	Framework (Application)	General Approach and Commentary
Banetta et al., 2001	Dynamic mathematical model of a generic cogeneration plant to evaluate the influence of small gas turbines in an interconnected electric network.	Ideal gas (IG) and perfect combustion (PX). Based on Simulink. They claim that the model may be utilised to represent plants with different characteristics and sizes.
Kikstra and Verkooijen, 2002	Gas turbine of only one component (helium) to design a control system.	Detailed model based on physical principles. No details are given concerning the independent variables.
Ghadimi et al., 2005	Software diagnosis to detecting faults like compressor fouling	PX without heat losses.
Gouws et al., 2006	To solve "practical problems" being experienced on a particular gas turbine combustor.	One-dim empirical model coupled with a commercial network solver (flows and pressures). The simplified model predicts the behaviour near a more detailed numerical model.
Jaber et al., 2007	To study the influence of different air cooling systems	IG with no gas components. Input data were the ambient conditions and the air cooling system configuration. Model validated against plant data.
Zhu and Frey, 2007	Desktop model for excel to represent a standard air Brayton cycle considering five gas components and combustion stoichiometrics with possibilities of oxygen excess.	Output turbine temperatures are empirical enthalpy functions. Inputs are variables like efficiencies, some pressure drops, temperatures, etc. This approach is not useful for a training simulator.
Kaproń and Wydra, 2008	Model based on IG expansion and compression to optimise the fuel consumption of a combined cycle power plant when the power has to be changed.	Model adjusts the gradient of the generated power change as a function of the weather forecast. Authors indicate that the results have to be compared with the real plant and that that main problem is to develop highly accurate plant model.
Rubechini et al., 2008	Four stage gas turbine model to predict the overall turbine performance based on three-dim Navier-Stokes equations. Coolant injections, cavity purge flows and leakage flows were included.	Four different gas models were used: three based on gas ideal behaviour with different specific heat Cp evaluation and one using real gas model with thermodynamic properties from tables. Combustion was not simulated. Authors say that a good model is needed to reproduce the correct thermodynamic behaviour of the fluid.
Chen et al., 2009	Model designed for optimisation running around the full load point ignoring combustion.	IG with detailed modelling of the flow through the equipment, heat transfer phenomena and basing the process on a temperature-entropy diagram.
Watanabe et al., 2010	To analyse the dynamical behaviour of industrial electrical power system.	Simulink used as platform. The governor system model and a simple machine infinite bus were considered. Model was validated against real data. No details of the combustor model are mentioned.

Table 1. Summary of previous published works.

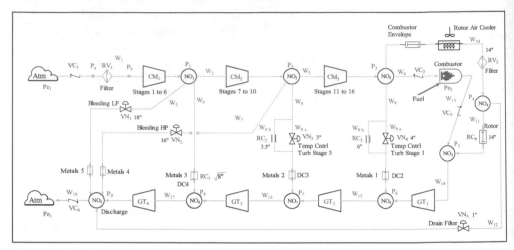

Fig. 2. Schematic diagram of the gas turbine-compressor-combustor system.

3. Description of the simulator

3.1 Hardware configuration

The simulation sessions are initiated and guided from the Instructor Console (IC). It consists in one PC (or simulation node) with two monitors. Two operators stations replicate the real control stations from the plant. Each of these stations has two monitors and one wall mounted screens (54") to manage and control the simulated power plant. There is an additional PC that is used as a backup and to make modifications to the software, process or control models and where they are tested and validated by an instructor before to install it in the simulator. Figure 3 presents a view of the simulator.

Fig. 3. GT simulator installed in the client facilities.

3.2 Software configuration

The simulator was developed under Windows XP and was programmed in Visual Studio Net, Fortran Intel, Flash and VisSim. The Simulation Environment (*MAS*) is copyright software of the IIE. A simplified conceptual diagram of the programs installed in the *MAS* is represented in Figure 4.

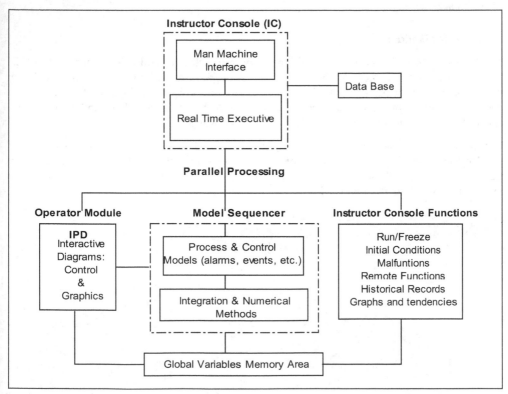

Fig. 4. Diagram of the Simulation Environment.

The platform, represented by the IC (with man-machine interfaces and a real time executive) has three main parts: the operator module, the model sequencer, and the IC functions. The communication between them is through a TCP/IP protocol. All the modules of the simulation environment except the Flash applications for the control interfaces are programmed with C# (Visual Studio).

The *MAS* is designed as a general tool for the GSACyS to develop simulators, it is very useful software that acts like a development tool and like the simulator man-machine interface. The *MAS* basically consist of four independent but coordinated applications: the real time executive, the operator module, the model sequencer and the instructor console.

The data base contains all the information required by the executive system. The tables contains information to support different process of the *MAS*, model sequencer, the number of times that a model is executed, Interactive Process Diagrams (IPD) for the instructor console and functions for the operators consoles. All the information is configurable for each application that uses the *MAS*.

The global variables memory area is a dynamic memory created when the simulation starts. This special area is formed from information of data base and contains the direction of global variables for both models process and control. This function allows to modify and to consult variables.

3.3 Operator Module

The Operator Module is a replica of the real operation station, and it is formed by the IPD (operation interfaces) that links the control and process models providing and receiving variables between them. Other functions of the module are to call the operation modes of each IPD component (valves, pumps, buttons, etc.); to refresh each IPD periodically; and to coordinate the sequence of events in the operation consoles.

A total of 20 control screens, including the index, and as many tendencies displays as needed are available for the operator. As an example, the blade path and exhaust temperatures control screen of the operator, a replica of the Distributed Control System (DCS), is shown in Figure 5.

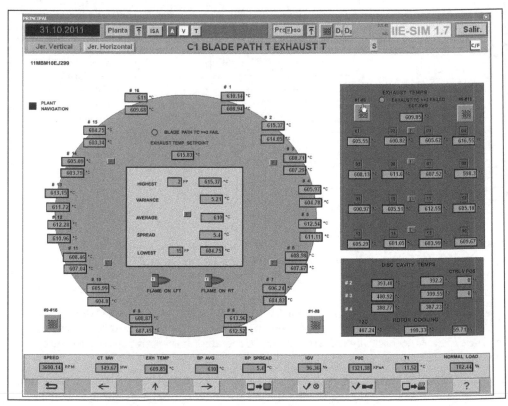

Fig. 5. Combustor blade path and exhaust temperatures control screens.

The flash movies have both, static and dynamic parts. The static parts are constituted by drawings of some particular control screens that are fixed on the screen. The dynamic parts are configured with graphic components stored in a library which are related to each one of the controlled equipments (pumps, valves, motors, etc.). These components have their own "moving" properties and they are employed during the simulation execution.

The main parts of the operator module are:

1. A main module (an "action-script" code) that loads the interactive process diagrams and gets information about the properties of the movie components.
2. A module to exchange information with the main module and with the executive system.
3. A module for the TCP/IP communications.

As a result of these functions, the student perceives his actions and simulator response in a very close way as it happens in the actual plant.

3.4 Models sequencer

The models sequencer coordinates the execution of mathematical models in a parallel scheme on a distributed architecture of PCs or with multi-core equipment. This module also has a group of methods to initialise variables belonging to the mathematical models. These values (initial conditions) represent a particular state of the plant and are located in the global variables memory area.

The models themselves and its mathematical treatment are explained in Section 5.1.

3.5 Instructor console functions

The Instructor console is presented in Figure 6.

Fig. 6. View of the instructor console.

The Instructor console has menus to activate different functions according the instructor necessities. The main functions are listed in this section.

Run/Freeze. The instructor may start or freeze a dynamic simulation session.

Initial conditions. The instructor may select an initial condition to initiate the simulation session, either from the general pre-selected list of initial conditions or from their own catalogue. Each instructor has access up to 100 initial conditions. Also in this function it is possible to create an initial condition or to erase an old one. As a instructional help, the simulator is capable of getting the automatic snapshooting function, every 15 seconds and may be configured up to a frequency of 10 minutes.

Simulation speed. From the beginning of the simulation session the simulator is executed in real time, but the instructor may execute the simulator up to ten times faster or ten times slower than real time.

Malfunctions. It is used to introduce, modify, or remove a simulated failure of plant equipment. For the case of the gas turbine simulator, there are 98 available malfunctions. Examples of malfunctions are: pumps trips, heat exchanger tubes breaking, heaters fouling, and valves obstructions. For the binary malfunctions (like trips), the instructor has the option to define its time delay and its permanence time. For analogical malfunctions (like percentage of a rupture), besides the mentioned time parameters, the instructor may define both intensity degree and evolution time.

Remote functions. The instructor has the option to simulate the operative actions not related with actions on the plant performed from the control screens. These actions are associated with the local actions made in the plant by auxiliary workers. Examples of them are: to open/close valves and to turn on/off pumps or fans. There are more than 200 of them and they may have time and intensity degree parameters as instructor options.

External parameters. The external conditions such as: atmospheric pressure and temperature (dry and wet bulb), voltage and frequency of the external system, fuel composition, gas fuel delivery pressure, etc. can be modified by the instructor.

Repetition. Simulation session may be repeated, exactly as operated by the trainees, including the trainer actions, as many times as the instructor considers it necessary. An action registration is the basis for this function.

Automatic exercises. This is a function that allows to create a file indicating a series of instructor functions to be automatically tripped along the simulation scenario in defined times.

Development tools. The simulator has implemented some others helpful tools to use during simulation session development, for example: to monitor and change on line the value of any selected list of global variables, tabulate any selected list of variables and plot them.

4. Real gas turbine power plant

The reference plant for this simulator is the unit 5 of "El Sauz", located in México. The nominal electric power is 150 MW and the plant is a pack generation unit Econopac 501F from Westinghouse. Generally speaking, this unit is formed for the gas turbine, the generator and auxiliary systems. It uses a system of low nitrogen oxide emissions DLN2 (Dry Low NOx - 2). The plant is designed to operate in an open cycle with natural gas.

Typically, a gas turbine power plant unit is based on the kinetic energy resulting from the expansion of compressed air and combustion gases. Resulting gases are directed over the turbine's blades, spinning the turbine, and mechanically powering the compressor and rotating the generator resulting in the production of electricity. After working in the turbine, the combustion gases are discharged directly into the atmosphere. In Figure 7, a simplified diagram of a gas turbine power plant process is shown.

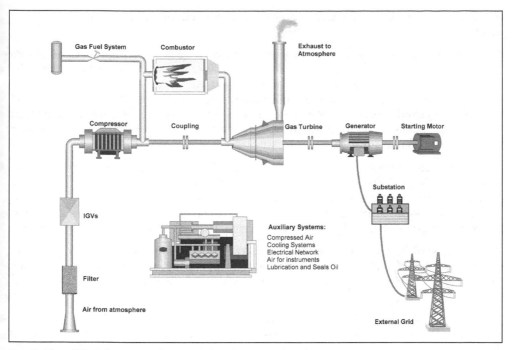

Fig. 7. Simplified diagram of a gas turbine power plant.

5. Developing of the gas turbine models

5.1 Execution of the models in the simulator

For an operators training simulation, the objective is to reproduce the behaviour of, at least, the reported variables in the plant control station in such a way the operator cannot distinguish between the real plant and the simulator performance considering both, the value of the parameters and their dynamics. The models were designed to work for all plant conditions, from 0% to 100% of load including all the possible transients that may be present during an operational session in the real plant. In order to accomplish with these objective, the "ANSI/ISA-S77.20-1993 Fossil-Fuel Power Plant Simulators Functional Requirements" norm was followed and adopted as design specification.

The gas turbine power plant was divided into a set of systems to be simulated, trying these to coincide with the real plant systems. A modelled system is a mathematical representation of the behaviour of the variables of the real system. The model of a simulated system (MSS) in the simulator responds to the operator's actions in the same way that the real system does, in tendency and time.

Each MSS is divided in algebraic and ordinary differential equations (AE and ODE, respectively). The AE may have to be solved simultaneously if necessary but they are independent of their ODE. For this work, different solution methods were adapted for linear and non-linear equations: Newton-Raphson, Gaussian elimination and bisection partition search. They are used depending on the characteristics of each particular model. The final equations set of each MSS are mathematically independent of the set of equations of any other MSS.

The ODE are solved numerically with Euler method or trapezoidal rule with one or two corrections. The integration step depends on the dynamic characteristic of the MSS. The determination of the proper integration method and integration step for a MSS is made running the model under a multistep multirate method to find out its "critical transient" (where the method reports more potential numerical instabilities). A numerical analysis involving calculation of Jacobian matrixes, eigenvectors and eigenvalues and its relation with the state variables is used.

For the simulator, an execution second (cycle) was divided into ten frames of 0.1 s. In this way, any model could be executed with an integration step of 0.1 s, 0.2 s, 0.5 s or 1 s. The sequence matrix indicates the execution precedence of the models defined by the developer. The sequence matrix indicates, besides the execution sequence, the integration method and integration step of each model, and the initial execution frame of each model for each cycle.

The definition of the sequence matrix was based on an algorithm that considers the number of inlet and outlet variables (causality variables) between the models to minimise the probability of mathematical instabilities. The sequence matrix data are actually in the simulator data base.

To exemplify the concept of sequence matrix, let say a simulator had four MSS: M1, M2, M3 and M4 and three integration methods. The sequence matrix as defined in Table 2, means that model M2 is the first in being executed in the first frame integrated with the method 1 and has a integration step of 0.1s, followed by model M1 with method 3 and integration step 0.2s; in the second frame model M4 with method 2 and integration step 0.2s is executed followed by model M2; in the third frame model M3 with method 1 and integration step 0.5s is executed, followed by models M2 and M1; and so on.

With the tenth frame executed, one simulation second has been finished. The executive program synchronises all the tasks in order to match the simulator time with real time. Typically, each control model is executed just after its controlled process model (and so the sequence matrix is constructed).

Model	Frames (10 frames is a second or cycle)									
	1	2	3	4	5	6	7	8	9	10
M3			1					1		
M4		2		2		2		2		2
M2	1	1	1	1	1	1	1	1	1	1
M1	3		3		3		3		3	

Table 2. Sequence matrix exemplified

5.2 General modelling methodology

The models of the simulator were implemented by the use of generic models, some times getting an improved version, and developing new models. In any case, a general procedure developed by the IIE is followed to obtain the final models as described in this section (Roldán-Villasana *et al.*, 2009).

Typically, the models are divided in process and control. However, there exist some common tasks to be done by any of the models:

- Data information of the process is obtained and classified: system description, system operation guidelines, thermal balances, operational curves of the equipment, operational plant data, etc.
- The information is analysed and a functional description of the MSS is made (conceptual model).
- Simplifications on the system are stated obtaining a simplified diagram (with the simulated equipment and their nomenclature). In the diagram the modelled equipment and measurements points are included and it is possible to identify, for example, the valves controlled automatically from those operated by the operator or locally.

5.2.1 Process models

The developing of the MSS of the process is described below, but regarding the general developing methodology, next points are observed:

- Main assumptions are stated and justified.
- The flow and pressure network configuration is obtained and the parameters of pumps, valves and fittings are obtained in a excel data sheet being this is a part of the final documentation. If it is modified, the updating of the simulator is automatically performed by reading the changes.
- Energy balances are programmed using the required generic models. Again, parameterisation with excel is performed. The calculations of energy balances consider possible sets of equations to be solved simultaneously.
- Other equipments are parameterised (boilers, condensers, etc.) using predefined excel sheets considering the simulator generic models or developing adequate models.

5.2.2 Control models

In the same way that the real plant, the control models acquire and process the actions realised by the operator on the control screens. The control models generate possible changes (perturbations) on the process models, for example, the demand of the valves to regulate some process variables like levels of tanks, pressures, temperatures, etc.

The DCS has most of the controls but some of them are local (out of the distributed control system). The graphics package VisSim was used for the development of the control models. Printed diagrams from the plant were the sources of the information in the format defined by the Scientific Apparatus Makers Association (SAMA).

The development of control models was carried out in five stages:

- Two hundred SAMA diagrams were reviewed and analysed to include all the control logic for the pre-start and start permissive; speed control; temperature monitoring; load control; and temperature control. As a result of this stage 21 control models were developed.

- Some equipments are not included in the DCS and the diagram or description of the local controls was sought. When the information was not available, the control diagrams were proposed and accorded with the client.
- A generic graphics library was integrated to create the analog and logic diagrams into the VisSim environment.
- The diagrams were transcribed into the VisSim program using the generic modules. Each SAMA control diagram was drawn into a single VisSim drawing.
- VisSim generates ANSI C code. Some adjustments to the code have to be done with a program developed by the IIE to translate the code to C#, which can run in the simulator environment. This application generates the variables definitions that are loaded into the simulator database (global variables, remote functions, malfunctions, etc.). The control models are kept in graphic form to facilitate the updates and future adjustments.

5.2.3 Coupling and adjustments

Once the process and controls models are ready, they have to be coupled and tested according next steps:

- Local tests are performed and necessary adjustments on the models are made, *i.e.* in the MAS the model is run alone (with its control) and changes in some input variables are made in order to evaluate the general response of the model.
- Coupling of the MSS and their controls are performed. All models are coupled into the MAS. In this case, the first model incorporated was the turbine, combustor and compressor model with its associated control. Then the fuel gas, then generator and electrical grid. Finally the models of the auxiliary systems, each with its control. The coupling order is an important factor and was done considering the best sequence to avoid as much as possible mathematical problems. An algorithm proposed and used previously in a simulator is to consider an execution sequence trying to minimise the retarded information (variables). To prove the successful integration of each model added, operation actions on the models were done.
- Fabric global test are made with the needed adjustments.
- Final acceptance tests are achieved by the final user according their own procedures.

5.3 Process modelling

5.3.1 Systems considered to be simulated

The systems that finally were simulated have all the information the instructor needs to train an operator and they may be controlled from the control screens, from the process itself, trough remote functions in the simulator, or from the DCS.

The systems included in the simulator are presented in Table 3. The simulator resulted with 13 process and 21 control models.

The combustor model includes the combustor blade path temperatures with 32 display values, the exhaust temperatures with 16 displays, the disc cavity temperatures with 8 displays, and emissions.

5.3.2 Models development

For the modelling, fundamental conservation principles were used considering a lumped parameters approach and widely available and accepted empirical relations. The lumped parameters approach simplifies the modelling of the behaviour of spatially distributed real systems into a topology consisting of few discrete entities that represent the behaviour of the distributed system (under certain assumptions). From a Mathematical point of view, the simplification reduces the state space of the system to a finite number, and the partial differential equations of the continuous (infinite-dimensional) time and space model of the physical system into ordinary differential equations with a finite number of parameters.

No.	Process Model	No.	Control Model	No.	Process Model	No.	Control Model
1	Fuel Gas	1	Pilot & stage control	10	Generator Cooling with Hydrogen	-	Temperature control
2	Gas Turbine and Compressors	2	Supervisory turbine	11	Electrical Network	12	Electrical system
		3	Rotor cooling control			13	Generator control
		4	Inlet Guide Valves control	12	Generator	14	VARS/Power Factor control
3	Turbine Metals, Temperatures, and Vibrations	5	Temperature control				
4	Combustor	6	Combustor	13	Performance Calculations (Heat Rate and Efficiencies)	-	-
		7	Combustor control		Several systems	15	Prestart and starting logic
		8	Flashback temperature monitoring			16	Auto-unload control
		9	System DLN control			17	Trip logic
5	Compressed Air	10	Air instrument control	-		18	Simulation/Test
6	Air for Instruments					19	Counter and timer logic
7	Lubrication Oil	11	Lube oil control			20	System alarms
8	Seals Oil						
9	Water-Ethylene Glycol Cooling	-	Temperature control			21	Graphic trends

Table 3. List of process and control models

Figure 8 shows the flow information of MSS. For each MSS, the independent variables (inlets) are associated with the actions of the operators like open or close a valve, trip a pump manually, etc., with the control signals, with the integration methods and with the information from other models.

A process model consists typically in the solution of the basic principles equations according the diagram shown in Figure 9. In this section the models related with the gas turbine, compressors and combustor are commented.

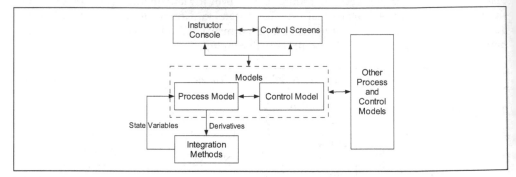

Fig. 8. Information flow of a model.

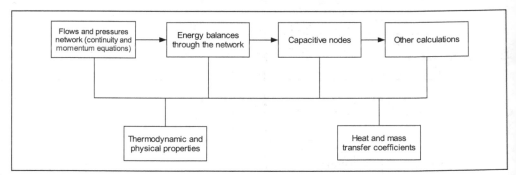

Fig. 9. Structure execution of a model.

Thermodynamic and physical properties

These properties are calculated as water (liquid and steam) and hydrocarbon mixtures thermodynamic properties and transport properties for water, steam and air.

For the water the thermodynamic properties were adjusted as a function of pressure (P) and enthalpy (h). The data source was the steam tables by Arnold (1967). The functions were adjusted by least square method. The application range of the functions is between 0.1 $psia$ and 4520 $psia$ for pressure, and -10 ^oC and 720 ^oC (equivalent to 0.18 BTU/lb and 1635 BTU/lb of enthalpy). The functions also calculate dTP/dP (being TP any thermodynamic property) for the saturation region and $\partial TP/\partial P$ and $\partial TP/\partial h$ for the subcooled liquid and superheated steam.

The hydrocarbon properties were applied for the gas fuel, air, and combustion products. Calculations are based in seven cubic state and corresponding state equations to predict the equilibrium liquid-steam and properties for pure fluids and mixtures containing non polar substances. The independent variables are pressure and temperature. The validity range is for low pressure to 80 bars. Twenty components were considered: nitrogen, oxygen, methane, ethane, propane, n-butane, i-butane, n-pentane, n-hexane, n-heptane, n-octane, n-nonane, n-decane, carbon dioxide, carbon monoxide, hydrogen sulphide, sulphur dioxide, nitrogen monoxide, nitrogen dioxide, and water.

The transport properties (viscosity, heat capacity, thermal expansion, and thermal conductivity) are calculated for liquid, steam and air with polynomial functions up to fourth degree.

Flows and Pressures Networks

This model simulates any hydraulic or gas network in order to know the values of the flows and the pressures along the system. The general approach to represent the network is considering that a hydraulic network is formed by accessories (fittings), nodes (junctions and splitters) and lines (or pipes). Accessories are those devices in lines that drop or increment the pressure and/or enthalpy of the fluid, like valves, pumps, filters, piping, turbines, heat exchangers and other fittings. A line links two nodes. A node may be internal or external. An external node is a point in the network where the pressure is known at any time, these nodes are sources or sinks of flow (inertial or capacitive nodes). An internal node is a junction or split of two or more lines.

The model is derived from the continuity equation on each of the nodes, considering all the inlet (i) and output (o) flowrates (w):

$$\Sigma w_i - \Sigma w_o = 0 \tag{1}$$

Also, the momentum equation may be applied on each accessory on the flow direction x, being ρ density, v velocity, g the gravity acceleration and τ viscous stress tensor:

$$\rho \frac{\partial v}{\partial t} = -\frac{\partial P}{\partial x} - \frac{\partial \tau_{xx}}{\partial x} - \rho v_x \frac{\partial v_x}{\partial x} + \rho g_x \tag{2}$$

Considering that the temporal and space acceleration terms are not significant, that the forces acting on the fluid are instantly balanced, a model may be stated integrating the equation along a stream:

$$\Delta P = -L \Delta \tau + \rho g \Delta z \tag{3}$$

The viscous stress tensor term may be estimated with empirical expressions for any accessory. For example, for a valve, the flowrate pressure drop (ΔP) relationship is:

$$w^2 = k' \rho \, Ap^\zeta \left(\Delta P + \rho g \Delta z \right) \tag{4}$$

Here, the flow resistance is function of the valve aperture Ap and a constant k' that depends on the valve itself (size, type, etc.). The exponent ζ represents the behaviour of a valve to simulate the relation between the aperture and the flow area. The aperture applies only for valves or may represent a variable resistance factor to the flow, for example when a filter is getting dirty. For fittings with constant resistance the term Ap^ζ does not exist. For a pump (or a compressor), this relationship may be expressed as:

$$\Delta P = k_1' \, w^2 + k_2' \, w \, \omega + k_3' \, \omega^2 - \rho \, g \, \Delta z \tag{5}$$

where ω is the pump speed and where k'_i are constants that fit the pump behaviour.

If it is considered that in a given moment the aperture, density, and speed are constant, both equations (4) and (5) may be written as:

$$w^2 = k_a \, \Delta P + k_b \tag{6}$$

Applying equation (1) on each node and equation (6) on each accessory a set of equations is obtained to be solved simultaneously. A more efficient way to get a solution is achieved if equation (6) is linearised. To exemplify, equation (6) is selected for the case of a pump with arbitrary numerical values (same result may be obtained for any other accessory). Figure 10 presents the quadratic curve of flowrate w on the x axis and ΔP on the y axis (dotted line). In the curve two straight lines may be defined as AB and BC and represent an approximation of the curve. The error is lower if more straight lines were "fitted" to the curve. In this case two straight lines are used to simplify the explanation, but the model allows for any number of them.

For a given flow w, the pressure drop may be approximated by the correspondent straight line (between two limit flows of this line). If there are two or more accessories connected in series and/or two or more lines in parallel are present, an equivalent equation may be stated:

$$w = C \, \Delta P + D \tag{7}$$

Substituting equation (7) on (1), for each flow stream, a linear equations system is obtained where pressures are the unknowns. The order of the equations system matrix is equal to the number of internal nodes of the network. Flows are calculated by equation (7) once the pressures were obtained.

The topology of a network may change, for instance, if a stream is "eliminated" or "augmented" for the network because a valve is opened and/or closed or pumps are turned on or off. The full topology is that theoretical presented if all streams allow flow through them. During a session of dynamic simulation a system may change its topology depending on the operator's actions. This means that the order of matrix associated to the equations that represent the system changes. To obtain a numerical solution of the model is convenient to count with a procedure that guarantees a solution in any case, *i.e.* avoiding the singularity problem. An algorithm was developed to detect the active topology in order to construct and solve only the equations related to the particular topology each integration time. The solution method is reported by Mendoza-Alegría *et al.* (2004). Figure 2 represents the flows and pressures network for the compressor and turbines.

Energy balances on internal nodes

Valves are considered isoenthalpic. The heat gained by the fluid due the pumping Δh_{pump} when goes through a turned on pump is:

$$\Delta h_{pump} = \left(\frac{\eta \, \Delta P}{\rho} \right) \tag{8}$$

where η is the efficiency.

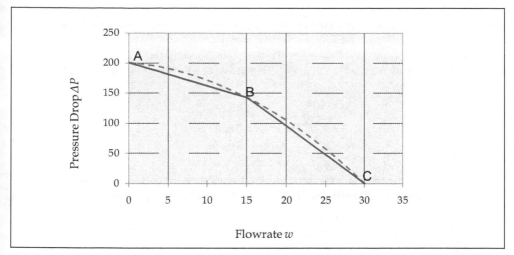

Fig. 10. Flowrate versus pressure drop linearization.

The exit properties for each turbine and compressor stage are calculated in a similar way, as an isoentropic expansion (or compression) and corrected with its efficiency. For example, the properties at the compressor's exhaust are calculated as an isentropic stage by a numerical Newton-Raphson method (i.e. the isoenthalpic exhaust temperature $T_o{}^*$ is calculated at the exhaust pressure P_o with the entropy inlet s_i):

$$f(T_o^*) = s_i(T_i, P_i, c) - s_o^*(T_o^*, P_o, c) = 0 \tag{9}$$

where c is the gas composition. Then, the leaving enthalpy is corrected with the efficiency of the compressor:

$$h_o = \frac{(h_o^* - h_i)}{\eta} + h_i \tag{10}$$

For the turbine it was considered that the work is produced instead of consumed. All other exhaust properties are computed with the real enthalpy and pressure. The efficiency η is a function of the flow through the compressor and turbine.

An energy balance on the flows and pressures network is made in the nodes where a temperature or enthalpy is required to be displayed or when a mixture of flowrates is made. In this case, the state variables are the enthalpy h, and the composition of all the components c_j is necessary in order to determinate all the variables of the node:

$$\frac{dh}{dt} = \frac{\sum w_i h_i - w_o h - q_{atm}}{m} \tag{11}$$

In this equation, m is the mass of the node, and q_{atm} the heat lost to the atmosphere. The subindex i represent the inlet conditions of the different flowstreams converging to the node. With the enthalpy and pressure it is possible to verify if the node works as a single or a two phase.

All the mass balances are automatically accomplished by the flows and pressures network solution, however, the concentration of the gas components must be considered through the network. The concentration of each species j is calculated as the fraction of the mass m_j divided by the total mass m in a node. The mass of each component is calculated by integrating next equation:

$$\frac{dm_j}{dt} = \sum w_i\, c_{i,j} - \sum w_o\, c_j \tag{12}$$

Combustor

To calculate the flame temperature, the oxidation equations for each one of the 20 potential components were stated. Some of the reactions for the combustion process are:

$$
\begin{aligned}
2n_{1,1}\,N_2 + 4n_{1,2}\,O_2 &\rightarrow 0m_{1,1}\,N_2 + 0m_{1,2}\,O_2 + 0m_{1,18}\,NO + 4m_{1,19}\,NO_2 \\
n_{2,1}\,O_2 &\rightarrow m_{2,2}\;\;O_2 \\
n_{3,1}\,H_2S + 1.5n_{3,2}\,O_2 &\rightarrow 0m_{3,2}\,O_2 + 0m_{3,3}\,H_2S + 1m_{3,16}\,H_2O + m_{3,20}\,SO_2 \\
n_{4,1}\,CO_2 &\rightarrow m_{4,4}\,CO_2 \\
n_{5,1}\,CH_4 + 2n_{5,2}\,O_2 &\rightarrow m_{5,4}\,CO_2 + 2m_{5,16}\,H_2O + 0m_{5,2}\,O_2 + 0m_{5,5}\,CH_4 \\
&\quad + 0m_{5,17}\,CO
\end{aligned}
$$

$$\bullet$$
$$\bullet$$
$$\bullet \tag{13}$$

$$
\begin{aligned}
10n_{15,1}\,C_{10}H_{22} + 155\,n_{15,2}\,O_2 &\rightarrow m_{15,4}\,CO_2 + m_{15,16}\,H_2O + 0m_{15,2}\,O_2 + 0m_{15,15}\,C_{10}H_{22} \\
&\quad + 0m_{15,17}\,CO \\
2n_{16,1}\,H_2O &\rightarrow 2m_{16,16}\,H_2O \\
n_{17,1}\,CO + 0.5n_{17,2}\,O_2 &\rightarrow m_{17,4}\,CO_2 + 0m_{17,17}\,CO + 0m_{17,2}\,O_2 \\
n_{18,1}\,NO + 0.5n_{18,2}\,O_2 &\rightarrow m_{18,19}\,NO_2 + 0m_{18,18}\,NO + 0m_{18,2}\,O_2 \\
n_{19,1}\,NO_2 &\rightarrow m_{19,19}\,NO_2 \\
n_{20,1}\,SO_2 &\rightarrow m_{20,20}\,SO_2
\end{aligned}
$$

The excess of oxygen may be calculated from the flowrates of the reactive components. The total combustion efficiency for each reaction i $\alpha_{i,1}$ is defined as the fraction of the theoretically amount of oxygen that is consumed for a total combustion: is 1 if a complete combustion reaction is hold, for example production of CO_2 and 0 if none of the products of a reaction is completely oxidised. The partial combustion efficiency $\alpha_{i,2}$ is defined as the fraction of the theoretically amount of oxygen that consumed for a partial combustion (is 1 if partial oxidised products are generated, for example production of CO and 0 if none of the products of a reaction is partially oxidised). The efficiencies are normally not constant and any function could be adjusted but considering that, to avoid imbalance problems, the restriction

$$\alpha_{i,1} + \alpha_{i,2} \leq 1 \tag{14}$$

must be satisfied at any moment.

The kinetics is not taken into account, but this approach considering these original two efficiencies, allows simulate the behaviour of the combustor. For this particular application of the combustor model, a linear function was defined for each efficiency, and for each reaction, based on the excess of oxygen (a thumb rule, but any equation could be defined).

As an example the equations to obtain the stoichiometric coefficients related with the nitrogen (and the oxygen in the nitrogen reaction) are:

$$N : 2n_{1,1} = 2m_{1,1} + m_{1,18} + m_{1,19}$$
$$O_2 : n_{1,2} = 0.5m_{1,18} + m_{1,19} \; ; m_{1,2} = 0.0$$
$$\text{If excess of oxygen} >= 200\% \quad a_{1,1} = 0.1 \; ; a_{1,2} = 0.0 \quad\quad (15)$$
$$\text{If excess of oxygen} = 100\% \quad a_{1,1} = 0.0 \; , a_{1,2} = 0.1$$
$$\text{If excess of oxygen} <= 50\% \quad a_{1,1} = 0.0 \; , a_{1,2} = 0.0$$

Here, a linear interpolation is used between the boundaries.

The reactive stoichiometric coefficients n are known variables except for the oxygen. Thus, for the nitrogen oxidation:

$$m_{1,19} = 2a_{1,1}n_{1,1}$$
$$m_{1,18} = 2a_{1,2}n_{1,1}$$
$$m_{1,1} = 0.5\left(2n_{1,1} - m_{1,18} - m_{1,19} \right) \quad\quad (16)$$
$$n_{1,2} = 0.5m_{1,18} + 1m_{1,19}$$

Similar equations may be stated for the oxidation reaction of each component and all the coefficients may be calculated. With the coefficient m, the concentration of the product is obtained. This formulation may be applied in a generic way for any quantity and any number of reactants.

An iterative process may be followed to find the flame temperature as a function of the amount of all the present species considering: the heat of combustion, calculated from the component's formation enthalpies; reactive and product sensitive heat; and heat losses by radiation and convection.

To develop the combustor model, it was conceptualised according Figure 11. Basically, the division considered a mixing node, the reaction and the capacitive node.

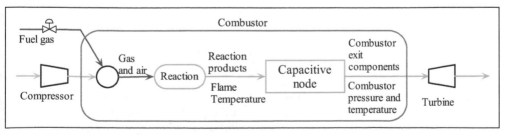

Fig. 11. Conceptual model of the combustor.

The mixing node just takes the air and fuel flowrates and concentrations to have single flowrate considering a perfect and instantaneous mixing. The reaction considers equations 13 to 16 to calculate the flame temperature, the flowrate and concentration of the products that enters into the combustor capacitive node, that has a volume and a concentration of the different species. Equation (12) applies for the mass of each component. The energy balance is done calculating the internal energy u in the node. The derivative of the internal energy is evaluated as:

$$\frac{du}{dt} = \frac{\sum w_i h_i - \sum w_o h - u \dfrac{dm}{dt} - q_c}{m} \tag{17}$$

The derivative of the total mass is the sum of the derivatives of each species as equation (12).

Here, the internal energy and density are known, this later by dividing the total mass and the volume of the node. The gas thermodynamic properties are a function of pressure and temperature, so all the properties are calculated with a double Newton-Raphson iterative method to have a solution. When two phases are present, some extra calculations are made to known the liquid and vapour volumes, but this is not treated here.

The combustor heat q_c is calculated as the sum of radiant and convective phenomena using appropriate correlations. The heat is absorbed by the combustor metal that is cooled with air flow in the rotor air cooler (see Figure 2). This heat is divided to be absorbed by different parts of the metal, like the blade paths presented in Figure 5. In each part, the modelling of the temperature may be represented by Figure 12.

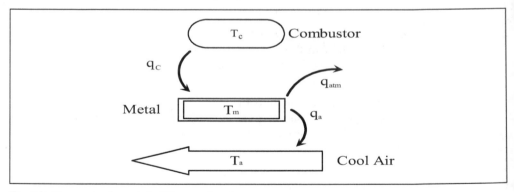

Fig. 12. Conceptual model of the combustor.

The temperature of the metal may be calculated by integration of next equation:

$$\frac{dT_m}{dt} = \frac{q_c - q_a - q_{atm}}{Cp_m\, m_m} \tag{18}$$

The metal heat capacity Cp_m is calculated with a polynomial function of temperature. Note that the term q_{atm} would not exist, depending on the particular position of each modelled metal. The exit air conditions changes due the absorbed heat. For these effects, Equation (11) is used.

Phenomena occurring in other equipments and systems like heat exchangers, electrical motors, generator, electrical network, tanks, etc., are used for other MSS but their formulations are not explained here.

6. Tests and results

The simulator testing was made with 16 detailed operation procedures elaborated by the costumer specialised personnel, namely the "Acceptance Simulator Test Procedures". The tests included all the normal operation range, from cold start conditions to full load, including the response under malfunctions and abnormal operation. In all cases the response satisfied the ANSI/ISA norm.

From the real plant an automatic start up procedure was documented (with a total of 302 variables obtained from the DCS). The tests presented here are the results of an automatic start-up followed by two malfunctions. During these transients no actions of the operator were allowed. In Table 4 a list of the events and the time they happen is presented.

Although they are not any more mentioned in this work, must be noted that all the expected alarms were presented in the precise time. Results include variables concerning the gas turbine power plant in general. The idea is demonstrate that the gas turbine model is capable to reproduce the real plant behaviour in all cases and that its response is adequate for training purposes. In all cases, plant data were available during the first 2780 s where the simulator results fit well.

Time Period (s)	Time Period (hh:mm:ss)	Event ID	Description
0 – 120	00:00:00 - 00:02:00	1	Simulation is initiated and plant stays in steady state for 5 s, then turbine is turned on in automatic mode and turbine roll-up initiates.
120 – 830	00:02:00 - 00:13:50	2	Combustor ignition is triggered and turbine reaches its nominal speed (3600 *rpm*).
830 – 2790	00:13:50 - 00:46:30	3	Initiates the electricity production up to nominal load (150 *MW*).
2790 – 2970	00:46:30 - 00:49:30	4	The plant is kept producing nominal load.
2970 – 3000	00:49:30 - 00:50:00	5	Progress malfunction of low efficiency in the combustor up to its final value (50% of severity).
3000 – 3500	00:50:00 - 00:58:20	6	Simulator runs without external changes, controls try to stabilise plant.
3500 – 3680	00:58:20 - 01:01:20	7	Progress malfunction of pressure loss in the delivery line of gas fuel up to its final value (4082 *kPa*).
3680 – 4200	01:01:20 - 01:10:00	8	Simulator runs without external changes, controls try to stabilise plant.

Table 4. Description of the programmed events for the test.

In Figure 13, the combustor pressure and the gas delivery pressure, an external parameter where the operator has no control, are presented on the y axis. The gas delivery pressure is measured in a pressure controlled header where the gas is stored from a duct that delivers the gas continuously.

Figure 13 shows how the gas delivery pressure changed with a ramp between the 3500 s and 3680 s. The behaviour of the combustor pressure may be explained as follows:

Event ID 4. The pressure stabilises.

Event ID 5. The pressure drops because the combustion is affected and the temperature in the combustor descents too.

Event ID 6. Pressure arises due the effect of the gas control valves that treats to keep the load of the plant by allowing more gas flowrate to the combustor.

Event ID 7. The pressure drops because the effect of the decreasing of the delivery pressure is greater than the effect of the gas control valves treating to keep the load. At about 3660 s the pressure increases and then descents because the aperture of the gas oscillate due the control algorithm.

Event ID 8. The pressure decreases lightly trying to reach a new, and final, steady state because the gas control valves tend to stabilise their aperture.

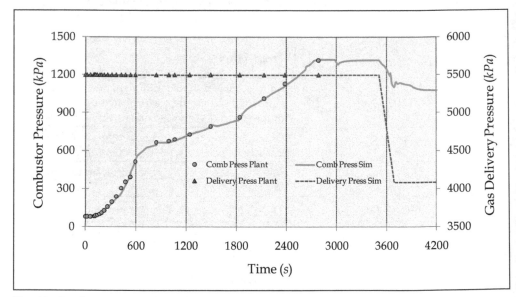

Fig. 13. Combustor pressure and gas delivery pressure

In Figure 14, the fuel gas flowrate and the apertures of two (A and C) of the gas control valve apertures are presented on the y axis. There exist four gas control valves, one for the pilots and three more that distribute the flow around the combustor inlet nozzles. All they open in a prefixed sequence and their function is to control the turbine speed and the produced electrical power.

The behaviour of the gas control valves dynamics and fuel gas flowrate during the malfunctions transients are:

Event ID 4. The apertures and the fuel gas flow remain stable.

Event ID 5. The apertures increase their value trying to compensate the load descending due the losing of electrical power because of the loss of combustion efficiency. The gas flowrate increases following the aperture of the gas control valves.

Event ID 6. The gas control valves adjust their value and control the load. The gas flowrate grows slightly according the gas control valves aperture.

Event ID 7. The gas control valves try to keep the load. At about 3660 s the gas control valves has a transient (open suddenly) but the control detects this abrupt change and limits the valve apertures producing a slight transient in the apertures. The gas flowrate descents because the gas control valves are not able to keep the load. The malfunction is too severe. At about 3660 s the gas flow follows the valves transient.

Event ID 8. The control algorithm allows decreasing the load and stabilise the apertures and the gas flow. The plant tends to a new steady state.

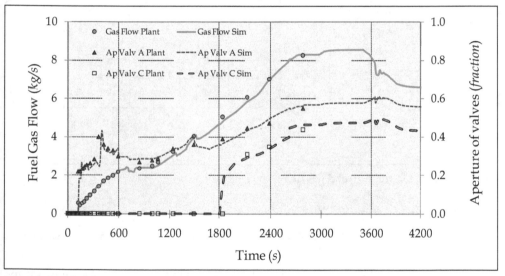

Fig. 14. Combustor pressure and gas delivery pressure

The exhaust temperature and the produced electrical charge by the generator are presented in Figure 15. The results may be summarised as follows:

Event ID 4. The exhaust temperature and load are in steady state.

Event ID 5. The exhaust temperature and load drop because the combustion is affected and the gas control valves do not respond immediately.

Event ID 6. The exhaust temperature goes up due the effect of the gas control valves that treats to keep the electrical power by opening and forcing more gas flowrate to the combustor. The load recovers its original nominal value due the control by the gas control valves.

Event ID 7. The exhaust temperature drops. The effect of the malfunction is more important than the effect of the gas control valves. At about 3660 s the exhaust temperature has a transient because the gas control valves aperture behaviour. The load drops because the malfunction of losing delivery pressure that cannot be nullified by the gas control valves. At about 3660 s the load oscillates due the gas control valves aperture behaviour.

Event ID 8. The exhaust temperature and load tend to stabilise according the gas control valves position.

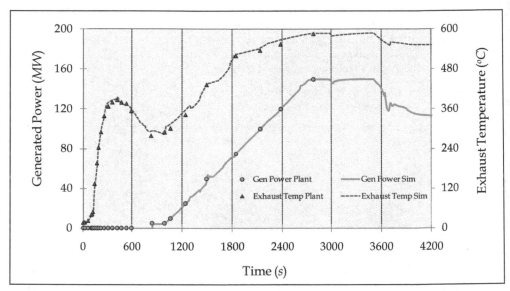

Fig. 15. Combustor pressure and gas delivery pressure

7. Conclusion

The convenience of the use of simulators for operators' training has been proved since long time ago.

The gas turbine simulator presented is a replica, high-fidelity, plant specific simulator for operators training.

The simulator was tested in all the operation range from cold start to 100% of load and fulfils the performance specified by the client, including a comparison of its results with plant data. It was tested and validated by the client itself (personnel having experience in the combined cycle power plant operation and in the use of simulators for qualification). This simulator presently is being used for training of CFE Turbo Gas power plant operators.

The project was designed and satisfactorily finished by following a methodology assuring the ending on time and with the goals properly accomplished. The *MAS* proved to be a developing platform as well as the final simulator platform.

The results confirmed that the simulator methodology and modelling approach is appropriate to develop a training device.

8. Acknowledgment

The GT simulator was successfully finished thanks to the endeavour, unconditional and professional work of researchers and technicians of the IIE as well as the good co-operation of operators, trainers and support personnel of the client. The simulator was developed with the financial support of CFE.

9. Acronyms

AE   - Algebraic Equations
ANSI   - American National Standards Institute
CFE   - Mexican Utility Company
CCS   - Combined Cycle Simulator
DCS   - Distributed Control System
DLN   - Dry Low NOx
GT   - Gas Turbine
HRSG   - Heat Recovery Steam Generation
IC   - Instructor Console
IG   - Ideal Gas
ISA   - International Standardization Association
IPD   - Interactive Process Diagrams
IIE   - Electrical Research Institute
MAS   - Simulation Environment
MSS   - Model of a Simulated System
ODE   - Ordinary Differential Equations
PX   - Perfect Combustion
PC   - Personal Computer
ProTRAX   - Commercial Power Plant Simulation Software
SAMA   - Scientific Apparatus Makers Association
GSACyS   - Advanced Training Systems and Simulation Department
TCP/IP   - Transmission Control Protocol/Internet Protocol
VisSim   - Visual Solutions Inc.

10. References

Banetta, S.; Ippolito, M.; Poli, D. & Possenti, A. (2001). A model of cogeneration plants based on small-size gas turbines, *16th International Conference and Exhibition on Electricity Distribution*, IEE Conf. Publ, Vol.4, No.482, ISSN: 0537-9989, January 18-21, 2001, Amsterdam, The Netherlands .

Chen, L.; Zhang, W. & Sun, F. (2009). Performance optimization for an open-cycle gas turbine power plant with a refrigeration cycle for compressor inlet air cooling. Part I: thermodynamic modelling, *Journal Power and Energy. Proc. IMechE*. Vol.223.

Epri (1993). Justification of Simulators for Fossil Fuel Power Plants, *Technical Report TR-102690*, EPRI, USA.

Fray, R.; Divakaruni M. (1995). Compact Simulators Can Improve Fossil Plant Operation, Power Engineering, Vol.99, No.1, pp.30-32, ISSN 0032-5961, United States.

Ghadimi, A.; Broomand, M, & Tousi, M. (2005). Thermodynamic model of a gas turbine for diagnostic software, Proceedings of the IASTED International Conference on Energy and Power Systems, pp. 32-36, ISSN 088986-548-5, April. 18-20, 2005, Krabi, Thailand.

Gouws, J.J.; Morris, R.M.; Vissr, J.A. (2006), Modelling of a gas turbine combustor using a network solver, South African Journal of Science, No.102, Nov/Dic., 2006.

Hoffman S. (1995). A new Era for Fossil Power Plant Simulators, Epri Journal, Vol.20, No.5, pp.20-27.

Hosseinpour, F.; Hajihosseini, H. (2009), Importance of simulation in manufacturing, World Academy of Science, Engineering and Technology, Vol.51, pp.285-288, March 2009, ISSN: 2070-3724.

Jaber, Q.M.; Jaber J.O. & Kawaldah M.A. (2007). Assessment of power augmentation from gas turbine power plants using different inlet air cooling system, Jordan Journal of Mechanical and Industrial Engineering. Vol.1, No.1, Sep. 2007, pp.7-17, ISSN 1995-6665.

Kaproń, H.; Wydra, M. (2008). Modelling of gas turbine based plants during power changes, The European Simulation and Modelling Conference Modelling and Simulation, pp.412-414, ISSN 0955301866-5.

Kikstra, J.F.; Verkooijen, A.H.M. (2002). Dynamic modelling of a cogenerating nuclear Gas turbine plant-part I: Modeling and Validation, Journal Engineering for Gas Turbines and Power, Vol. 124, July 2002, pp.725-733.

Mendoza ,Y.; Roldán-Villasana, E.J.; Galindo, I. & Romero, J. (2004). Methodology to adapt the feedwater and condensate system using a flow and pressure generic model for the Laguna Verde nuclear power plant simulator, Proceedings of Summer Computer Simulation Conference, The Society for Modeling and Simulation International, pp. 123-128, ISBN:1-56555-283-0, July 25-29, 2004, San Jose California, USA.

Roldán-Villasana, E.J.; Cardoso, M.J.; Mendoza-Alegría, Y. (2009), Modeling Methodology for Operators Training Full Scope Simulators Applied in Models of a Gas-Turbine Power Plant, Memorias del 9o. Congreso Interamericano de Computación Aplicada a la Industria de Procesos, pp. 61-66, 25 al 28 de agosto, 2009, Montevideo, Uruguay.

Rubechini, F.; Marconcini, M.; Arnone, A.; Maritano, M.& Cecchi, S. (2008). The impact of gas modeling in the numerical analysis of a multistage gas turbine, Journal of Turbomachinary, Vol.130, pp.021022-1 – 021022-7, April 2008.

Vieira, L.; Matt, C.; Guedes, V.; Cruz, M. & Castelloes F. (2008). Optimization of the operation of a complex combined- cycle cogeneration plant using a professional process simulator, Proceedings of IMECE2008, ASME International Mechanical Engineering Congress and Exposition, pp.787-796, October 31-November 6, 2008, Boston, Massachusetts, USA.

Watanabe, M.;Ueno,Y.;Mitani,Y.; Iki,H.; Uriu,Y. & Urano, Y. (2010). Developer of a dynamical model for customer´s gas turbine generator in industrial power systems, 2nd IEEE International Conference on Power and Energy, (PECon 08). December 1-3, pp.514-519, 2008, Johor Baharu, Malaysia.

Zhu, Y., Frey, H.C. (2007). Simplified performance model of gas turbine combined cycle systems, Journal of Energy Engineering. Vol.133, No.2, Jun 2007, pp.82-90, ISSN 0733-9402.

Permissions

The contributors of this book come from diverse backgrounds, making this book a truly international effort. This book will bring forth new frontiers with its revolutionizing research information and detailed analysis of the nascent developments around the world.

We would like to thank Konstantin Volkov, for lending his expertise to make the book truly unique. He has played a crucial role in the development of this book. Without his invaluable contribution this book wouldn't have been possible. He has made vital efforts to compile up to date information on the varied aspects of this subject to make this book a valuable addition to the collection of many professionals and students.

This book was conceptualized with the vision of imparting up-to-date information and advanced data in this field. To ensure the same, a matchless editorial board was set up. Every individual on the board went through rigorous rounds of assessment to prove their worth. After which they invested a large part of their time researching and compiling the most relevant data for our readers. Conferences and sessions were held from time to time between the editorial board and the contributing authors to present the data in the most comprehensible form. The editorial team has worked tirelessly to provide valuable and valid information to help people across the globe.

Every chapter published in this book has been scrutinized by our experts. Their significance has been extensively debated. The topics covered herein carry significant findings which will fuel the growth of the discipline. They may even be implemented as practical applications or may be referred to as a beginning point for another development. Chapters in this book were first published by InTech; hereby published with permission under the Creative Commons Attribution License or equivalent.

The editorial board has been involved in producing this book since its inception. They have spent rigorous hours researching and exploring the diverse topics which have resulted in the successful publishing of this book. They have passed on their knowledge of decades through this book. To expedite this challenging task, the publisher supported the team at every step. A small team of assistant editors was also appointed to further simplify the editing procedure and attain best results for the readers.

Our editorial team has been hand-picked from every corner of the world. Their multi-ethnicity adds dynamic inputs to the discussions which result in innovative outcomes. These outcomes are then further discussed with the researchers and contributors who give their valuable feedback and opinion regarding the same. The feedback is then collaborated with the researches and they are edited in a comprehensive manner to aid the understanding of the subject.

Apart from the editorial board, the designing team has also invested a significant amount of their time in understanding the subject and creating the most relevant covers. They scrutinized every image to scout for the most suitable representation of the subject and create an appropriate cover for the book.

The publishing team has been involved in this book since its early stages. They were actively engaged in every process, be it collecting the data, connecting with the contributors or procuring relevant information. The team has been an ardent support to the editorial, designing and production team. Their endless efforts to recruit the best for this project, has resulted in the accomplishment of this book. They are a veteran in the field of academics and their pool of knowledge is as vast as their experience in printing. Their expertise and guidance has proved useful at every step. Their uncompromising quality standards have made this book an exceptional effort. Their encouragement from time to time has been an inspiration for everyone.

The publisher and the editorial board hope that this book will prove to be a valuable piece of knowledge for researchers, students, practitioners and scholars across the globe.

List of Contributors

Konstantin Volkov
School of Mechanical and Automotive Engineering, Faculty of Science, Engineering and Computing, Kingston University, London, UK

Roberto Capata
Department of Mechanical and Aerospace Engineering, University of Roma 1, Faculty of Engineering, Roma, Italy

Rahim K. Jassim
Department of Mechanical Engineering Technology, Yanbu Industrial College, Yanbu Industrial City, Saudi Arabia

Majed M. Alhazmy and Galal M. Zaki
Department of Thermal Engineering and Desalination Technology, King Abdulaziz University, Jeddah, Saudi Arabia

I. Gurrappa and A. K. Gogia
Defence Metallurgical Research Laboratory, Kanchanbagh PO, Hyderabad, India

I. V. S. Yashwanth
M.V.S.R. Engineering College, Nadargul, Hyderabad, India

Jarosław Milewski, Krzysztof Badyda and Andrzej Miller
Institute of Heat Engineering at Warsaw University of Technology, Poland

Igor Loboda
National Polytechnic Institute, Mexico

Roberto Carapellucci and Lorena Giordano
Department of Mechanical, Energy and Management Engineering, University of L'Aquila, Italy

Edgardo J. Roldán-Villasana and Yadira Mendoza-Alegría
Instituto de Investigaciones Eléctricas, Gerencia de Simulación, México